A Mathematically Coherent Quantum Gravity

A Mathematically Coherent Quantum Gravity

James Moffat
Kings College, University of Aberdeen, Aberdeen, UK

IOP Publishing, Bristol, UK

ISBN 978-0-7503-2580-6 (ebook)
ISBN 978-0-7503-2578-3 (print)
ISBN 978-0-7503-2581-3 (myPrint)
ISBN 978-0-7503-2579-0 (mobi)

DOI 10.1088/978-0-7503-2580-6

Version: 20200901

IOP ebooks

British Library Cataloguing-in-Publication Data: A catalogue record for this book is available from the British Library.

Published by IOP Publishing, wholly owned by The Institute of Physics, London

IOP Publishing, Temple Circus, Temple Way, Bristol, BS1 6HG, UK

US Office: IOP Publishing, Inc., 190 North Independence Mall West, Suite 601, Philadelphia, PA 19106, USA

The cover design is based on a painting by the author. It is called ‘Logos’ which is Greek for Logic.

I dedicate this book jointly to Frank Bonsall and John Ringrose.

Contents

Preface: why you should read this book

In the very last lines of *The Meaning of Relativity*[1] at the end of appendix 2, Albert Einstein writes;

One can give good reasons why reality cannot at all be represented by a continuous field. From the quantum phenomena it appears to follow with certainty that a finite system with finite energy can be completely described by a finite set of numbers (quantum numbers). This does not seem to be in accordance with continuum theory and must lead to an attempt to find a purely algebraic theory for the description of reality. But nobody knows how to obtain the basis of such a theory.

This book is an algebraic theory for the description of reality, based on the development of algebras of quantum operators on Hilbert space by John von Neumann.

The development of a successful theory of quantum gravity in the context of the early universe is the key next step in theoretical physics. This book takes that step by describing a coherent mathematical framework for both the evolution of discrete space–time and the quantum graviton in the Planck regime. In doing so it successfully blends developments in both loop quantum gravity and superstring theory with data from the Large Hadron Collider Run 2 at 13 TeV.

In perturbative forms of quantum gravity, it is typically assumed that an extension of Dirac's worldview applies. A background space–time exists. A stage is provided, and the role of physics is to correctly model the dynamical interactions of the actors on that stage. However, a fundamental theory must model the evolution of the stage as well as the players. In terms of the early evolution of space–time during the 'Planck era' or 'Planck regime,' corresponding to some small multiple of the Planck time, the key relativistic property was spin.

- In the approach laid out here, I describe the algebraic language of von Neumann algebras (chapter 1).
- The discrete steps across a network of entangled spins forming a quantum computation during the Planck regime are the basis of a path through space–time (chapter 2).
- These pathways are the foundation on which more elaborate constructs such as particles can be created (chapter 3).
- And quantum fields (chapter 4) can be established. I also show in chapter 4, that when the approximation between quantum and classical ergodic theory is sufficiently close to allow the application of a Carnot cycle to the entropy of the surface area of a black hole, then we have, embedded in our theory as an approximation, Einstein's field equations for gravity.

[1] Einstein A 1960 *The Meaning of Relativity* 6th edition (London: Methuen).

- Finally, in chapter 5, I discuss a number of areas for future mathematical exploration.

In more detail

At the beginning of the Big Bang Universe, Einstein's theory of the gravitational field (general relativity) breaks down since it implies that the curvature is infinite, which does not make sense. In this book I identify the key issue as being that, like a less favoured aunt, general relativity brings with it far too much baggage. I redefine the underlying notions of space–time as simpler and sparser properties of a discrete quantum theory. This I do using ideas drawn from the logic of quantum computation. A trajectory in space–time becomes a quantum algorithm corresponding to a set of logical steps moving a quantum state across a network of entangled spin states—a process known in quantum computing as quantum teleportation. In this way I produce a simpler, sparser, discrete set of notions about gravity. Gravity becomes a set of waymarks which tell you how to navigate the discrete set of paths corresponding to a quantum computational network. These ideas are then compatible with our most modern definition of a quantum state[2] as a normalised, positive element of the dual space of an algebra of quantum observables. This is the essence of chapters 1 and 2. In terms of a coherent quantum field theory, the implications are as follows:

(a) Each local algebra *O*(*D*) should correspond to the algebra of sections of a principal fibre bundle with base space a finite and bounded subset of space–time[3]. The algebraic operations of addition and multiplication are assumed defined fibrewise for the algebra of sections.

(b) There is a limit on the resolution of the space–time variables and a corresponding limit to the energy available to the system—i.e. an upper energy or 'UV cut-off'.

In chapter 3, I show how particles, including the graviton, emerge in such a theory by exploiting the mathematics of heterotic string theory projected from 10 dimensions down to our four-dimensional space–time. Chapter 4 is the final development of our theory. Here I create a new physics and mathematics corresponding to a new way of thinking about what I call local algebras. These are algebras of quantum fields defined over a compact subset of space–time and subject to the constraints of symmetry invariance and symmetry-breaking processes. One motivation for this development was Ashtekar's[4] interest in diffeomorphism invariance. I have, in chapter 4, building on the developments in chapters 1–3, shown that such a quantum field can be identified with a local algebra by rolling up each such quantum field into a local quantum operator. The quantum field then becomes identified both as a section through the resultant fibre bundle and as an

[2] Sakai S 1971 *C*-Algebras and W*-Algebras* (Berlin: Springer).

[3] Araki H 2009 *Mathematical Theory of Quantum Fields* (Oxford: Oxford University Press).

[4] See for example Ashtekar A *et al* 1995 Quantization of diffeomorphism invariant theories of connections with local degrees of freedom *J. Math. Phys.* **36** 11).

element of a local W*-algebra[5]. I have also shown that, for such a construct, the relativistic quantum algebraic structure of the algebra can be identified, if G-type III, with the type III structure of the cross-product. I have created a new approach to the understanding of Yang–Mills quantum fields. This leads on to the development of quantum ergodic theory as a non-commutative extension of classical ergodic theory. We are then, surprisingly, able to demonstrate that when this quantum ergodic theory is sufficiently close to the classical case to allow the application of a Carnot cycle to the entropy of the surface area of a black hole, then we have, embedded in our theory as an approximation, Einstein's field equations for gravity.

Finally, in chapter 5, I introduce a number of developments of the theory for mathematics including;

- the Clay Mathematics Institute Millennial Question concerning Yang–Mills quantum theory and the mass gap.
- A discussion of the application of Mackey theory[6] to the development of sections through a fibre bundle.
- The quantum physics of a discrete closed string in the Planck regime giving rise to a 'stack' of phonon-like frequencies, and addressing the following problem; is it possible to describe an initial discrete form of the closed relativistic string during the Planck regime which then evolves into the form we know as a smooth approximation on a four-manifold, as the Universe expands and cools.

This book had its genesis in my PhD research during the period 1971–4 on the subject *Groups of Automorphisms of Operator Algebras* funded by a research scholarship provided by the Carnegie Trust. This was arranged to take place at Newcastle University by Professors Frank Bonsall (University of Edinburgh) and Professor John Ringrose (Newcastle). During my undergraduate years at Edinburgh (1967–71, Summa Cum Laude, Napier Medallist), Professor Bonsall introduced me to this amazing area of Mathematics. Professor Ringrose supervised my PhD and the resultant thesis is now available online in scanned form as part of the Newcastle thesis archive at https://theses.ncl.ac.uk. I then had to wait for around 20 years for this 'beautiful and difficult' subject to become Physics—as amusingly shown on the front cover of the excellent two volume work *Quantum Fields and Strings: A Course for Mathematicians* (published by the AMS and Princeton Institute for Advanced Study in 1999).

[5] In this book, I make no distinction between a von Neumann algebra situated on a Hilbert space and an abstract W*-algebra. Any W*-algebra, through the GNS construct, can be shown to be isomorphic to such a von Neumann algebra.

[6] See Loomis L 1953 *Abstract Harmonic Analysis* (New York: Van Nostrand).

Acknowledgements

The Author gladly acknowledges the help of Mr Andrew Partridge in formatting of the text.

Author biography

James Moffat

Professor James Moffat C Math FIMA FOR has an honorary chair in Physics at the University of Aberdeen (Kings College) where he is part of the Quantum Gravity and Advanced Engineering research group. His www.Researchgate.net publication impact score puts him in the top 20% of scientists globally; based on over 100 peer reviewed publications and several books, mostly published in the USA. Prior to that he was one of the most senior scientists advising the UK government on Defence and Security and has a global reputation for mathematical modelling. Winner of the President's Medal of the Operational Research Society (The 'Nobel Medal in Analytics') the Napier medal in mathematics and the Einstein Medal of the US Marine Corps. He is an advisor to the Parliamentary and Scientific Committee of the UK Parliament.

Symbols

When a new idea or definition is introduced; it is expressed in italic text.

IOP Publishing

A Mathematically Coherent Quantum Gravity

James Moffat

Chapter 1

Quantum theory

1.1 Basic notions

Quantum theory is a guess at how inputs to an experiment at the quantum level relate to outputs. These inputs and outputs are real numbers read on displays in the domain of classical mechanics, and, following the ideas of N Bohr, assumptions concerning experimental outcomes are part of the theory. The experiment can be real or conjectural, leading to a test of validation or predictive power. Good guesses are those which are consistent with all the existing experimental data (the facts) and have high predictive power, within their domain of application. This approach helps us to address issues such as 'don't ask questions just calculate'. Our currently accepted theory is the simplest of the good guesses, because the simplest explanation consistent with the facts is, we assume, nearest to the truth. This assumption is called the Principle of Parsimony; a stronger form of 'Occam's Razor' [1]. This book will be a journey in which we (that is you the reader and I the author) will travel together through an undiscovered land in order to reach a 'purely algebraic theory for the description of reality'; something which Einstein could not achieve [2].

From these preliminary remarks I can assume that a good theory of quantum gravity will have the following characteristics;

- It will have an elegant mathematical structure, since this is the simplest way of expressing its logic.
- It will be a theory based on the transformation of existing information into new information.
- It will be a black box model, with full access to its inputs and outputs but potentially incomplete information about why it works, since I do not require this of a good guess, model or theory. All I require is that for a given set of inputs, the outputs are consistent with the facts, within its domain of application.

doi:10.1088/978-0-7503-2580-6ch1

From the point of view of linear operator theory (the Heisenberg–Dirac viewpoint), we can interpret quantum theory as follows.

A system observable capable of experimental measurement such as the total energy corresponds to an element of the set $\mathcal{B}(\mathcal{H})$ of all bounded linear operators acting on a Hilbert space $\mathcal{H}$. Since a given quantum system has finite energy this space is either finite dimensional, if the energy is specified precisely, or separable when it is more convenient to allow the energy not to be specified. An example of the latter case would be a Fock space. Following Dirac [3], all of the system observables of interest are self-adjoint, have discrete real eigenvalues as measured values, and form a subset of $\mathcal{B}(\mathcal{H})$.

It will be convenient for us also to consider more general topological subsets of $\mathcal{B}(\mathcal{H})$ known as von Neuman algebras [4], denoted as $\mathcal{M}$. The corresponding operator topologies on $\mathcal{M}$—the weak, strong and ultraweak—will be discussed in detail later in this chapter.

A *pure quantum state* is an extreme element f of the closed convex subset of the dual space $\mathcal{M}^*$ generated by the set $\boldsymbol{S}$ of all states for the weak*-topology. This topology is the weak topology on the dual space induced by the Gelfand transform embedding

$$A \to \hat{A} \in \mathcal{M}^{**}; \quad \hat{A}(f) = f(A).$$

It is thus equivalent to the convergence criterion $\{f_n\} \to f \Leftrightarrow f_n(A) \to f(A);\ A \in \mathcal{M}$, induced by the seminorms $\{p_A;\ p_A(f)\} = |f(A)|$.

- When the Hilbert space is finite dimensional, a pure quantum state is equivalent to a normalized Dirac ket $|x\rangle$ of the Hilbert space [4]. It is an element of the dual space $\mathcal{M}^*$ of $\mathcal{M}$ via its action defined as;

$$\omega_x\colon A \to \langle x, Ax\rangle;\ A \in \mathcal{M}.$$

1.2 Non-relativistic quantum theory

For our purposes here I define a wave to be a disturbance which propagates with constant velocity in a given direction. This disturbance is assumed to have a constant profile which does not decay over time. If $F(x(t),t)$ is the profile of such a disturbance, travelling along the x-axis, at location $x(t)$ at time t, then from our definition of a wave it follows that the profile at time t, $F(x(t),t)$, is the same as the profile at time zero; $F(x(0),0)$, which equals $F(x(t){-}ct,0)$ where c is the speed with which the disturbance moves. Setting $\tau = x(t){-}ct$ leads to the profile function $G(\tau) = F(\tau, 0)$ and the wave equation;

$$\frac{\partial F(x(t),\,t)}{\partial t} = \frac{\partial G(\tau)}{\partial \tau}\frac{\partial \tau}{\partial t} = -c\frac{\partial G(\tau)}{\partial \tau}.$$

Thus

$$\frac{\partial^2 F(x(t), t)}{\partial t^2} = c^2 \frac{\partial^2 G(\tau)}{\partial \tau^2}.$$

Similarly

$$\frac{\partial^2 F(x(t), t)}{\partial x^2} = \frac{\partial^2 G(\tau)}{\partial \tau^2} = \frac{1}{c^2}\frac{\partial^2 F(x(t), t)}{\partial t^2}.$$

Periodic solutions of this wave equation are of the form $F(x(t), t) = \exp i(\pm ax \pm st)$.

Inserting this into the wave equation it follows that $\frac{s^2}{c^2} = a^2$. Thus, such a periodic plane wave has the typical form $F(x(t), t) = \exp i\left(\pm\frac{s}{c}x + st\right)$.

Setting the time to zero in this expression shows that the initial disturbance is a sinusoidal plane wave which then travels at constant speed along the x-axis. Let us therefore assume a periodic plane wave solution with the form;

$$F(x(t), t) = \exp 2\pi i\left(\frac{x}{\lambda} - \nu t\right)$$

where ν is the frequency of this wave, λ the wavelength. Substituting the de Broglie relations for the energy E and momentum p of a free particle into this wave amplitude function gives;

$$E = h\nu; \quad p = \frac{h}{\lambda}$$

$$\psi(x, t) = \exp\frac{2\pi i}{h}(px - Et).$$

This is the most basic example possible of a Schrödinger 'matter wave' function. It corresponds to a free running particle moving in the x-direction with energy E and momentum p.

The set of all such reasonably behaved wave functions for a single particle forms a space called a linear space or vector space. Given two wave functions $\psi_1(x, t); \psi_2(x, t)$ in this linear space I can also form the inner product;

$$\langle\psi_1(t)|\psi_2(t)\rangle = \int \psi_1^*(x, t)\psi_2(x, t)dx.$$

The ability to form such physical inner products, linear in the second term and conjugate linear in the first, makes the linear space of wave functions into a pre-Hilbert space. This space is a metric space relative to the norm topology. The completion is thus a closed set denoted $\mathcal{H}$. These wave functions relate to probabilities, thus I require that they are normalised;

$$\langle\psi(t)|\psi(t)\rangle = \int \psi^*(x, t)\psi(x, t)dx = 1$$

for all times t.

1.2.1 The measurement process

We now describe the process of measuring the energy of the wave function state $\psi(x, t)$. Note that if I differentiate this wave function in terms of time t then it pulls out the value E for the energy from the right-hand side, allowing it to be measured;

$$i\hbar\frac{\partial}{\partial t}\psi(x, t) = i\hbar\frac{\partial}{\partial t}\left\{\exp\frac{2\pi i}{h}(px - Et)\right\} = E\psi(x, t).$$

In other words, the linear operator $i\hbar\frac{\partial}{\partial t}$ has E as an eigenvalue for the energy eigenvector $\psi(x,t)$. Similarly, the linear operator $-i\hbar\frac{\partial}{\partial x}$ has momentum eigenvalue p;

$$-i\hbar\frac{\partial}{\partial x}\psi(x, t) = -i\hbar\frac{\partial}{\partial x}\left\{\exp\frac{2\pi i}{h}(px - Et)\right\} = p\psi(x, t).$$

Let $\hat{H} = i\hbar\frac{\partial}{\partial t}$ so that for our basic wave function ψ *I* have $\hat{H}\psi = E\psi$ then;

- $\hat{H}$ is a self-adjoint linear operator since its eigenvalues are real numbers.
- Any reasonable wave function can be expressed as a weighted sum of these basic wavefunctions.
- Our black box assumptions about quantum theory imply that this linear operator represents the observable or measurable energy of a non-relativistic quantum state.

Taking these together means that if ψ_j are the basic eigenfunctions corresponding to energy measurements E_j and ψ is a general wave function of the particle, then we can measure the energy of the quantum state defined by this general wave function as follows;

We have complex numbers α_j such that;

$$\psi = \sum_j \alpha_j \psi_j$$
$$\hat{H}\psi = \sum_j \alpha_j \hat{H}\psi_j = \sum_j \alpha_j E_j \psi_j.$$

Now instead of one energy level there is a weighted combination of several of them; a *mixed state*. Because we insisted that all wave function states, both the basic pure states ψ_j and the mixed state ψ are normalized, this implies that $\sum|\alpha_j|^2 = \langle\psi, \psi\rangle = 1$. Following Dirac, we interpret the $|\alpha_j|^2$ as probabilities. Measurement of the energy of a general mixed state then happens as follows.

With probability $|\alpha_j|^2$ we project down onto the basic eigenfunction Ψ_j and then the measured value of the energy is the corresponding eigenvalue E_j. From our earlier discussion, we can interpret this as follows;

- A quantum state is a normalized element $|x\rangle$ of a Hilbert space which is, in turn, either finite or separable.
- The inner product $\langle x, y\rangle$ represents the (complex) amplitude of the transition from state $|x\rangle$ to $|y\rangle$.

- A system observable such as energy corresponds to a bounded linear operator acting on this Hilbert space.
- Measurements correspond to discrete eigenvalues of this linear operator.

For example, a photon of light is the energy emitted when an electron falls from a higher energy level (energy eigenstate and eigenvalue) to a lower one. If the levels were not quantized this would produce a continuous blur of spectral values, in contradiction to what is actually seen. *More generally, Bohr's quantum approach, producing discrete energy eigenstates and eigenvalues, explains the stability of matter over time.*

1.2.2 The Schrödinger equation

For our free particle with momentum $p(x,t)$, we have just kinetic energy. The particle is not trapped. The classical (non-quantum) version of the Hamiltonian in this case is

$$H(x, t) = \frac{1}{2m}p(x, t)^2.$$

We know that a quantum system has to look like the classical non-quantum system at a larger scale as we pull back from the detail. This is called the correspondence principle. We thus guess that the quantum version of the Hamiltonian is of the form;

$$\hat{H}(x, t) = \frac{1}{2m}\hat{p}(x, t)^2.$$

Substituting in the linear operators for energy and momentum gives;

$$i\hbar\frac{\partial}{\partial t}\psi(x, t) = -\frac{1}{2m}\hbar^2\frac{\partial^2}{\partial x^2}\psi(x, t).$$

This is *Schrödinger's equation* for a free particle. In the general case we also have a potential function of the form $V(x)$, normally depending only on position. This is represented by the linear operator $L_{V(x)}$: $\psi \to V(x)\psi$.

The Hamiltonian in this case is then the sum of the kinetic and potential energies;

$$\hat{H}(x, t)\psi(x, t) = -\frac{\hbar^2}{2m}\frac{\partial^2}{\partial x^2}\psi(x, t) + V(x)\psi(x, t).$$

Substituting for $\hat{H}$ gives the general Schrödinger equation and the energy eigenvalue equation;

$$i\hbar\frac{\partial}{\partial t}\psi(x, t) = -\frac{\hbar^2}{2m}\frac{\partial^2}{\partial x^2}\psi(x, t) + V(x)\psi(x, t)$$

$$-\frac{\hbar^2}{2m}\frac{\partial^2}{\partial x^2}\psi(x, t) + V(x)\psi(x, t) = E\psi(x, t).$$

In solving these equations, the location and time components of the wave function are separated out. Thus, this theory is inherently non-relativistic.

1.3 Exploiting quantum theory embedded in classical gravity

The first quantum technology revolution (Quantum 1.0) was based on the invention of the transistor during the period 1925–48. The idea was a practical implementation of Dirac's generalised 'theory of quantum holes' in a stack of electron energy levels corresponding to positive quasi-particles. This forms the basis of our understanding of the quantum behaviour of solid-state semiconductors and layered semiconductors known as 'transistors' [5]. An 'n–p' junction between a silicon wafer doped with excess electrons (n-type) and a wafer with few electrons (p-type); equivalently with an excess of positively charged 'holes'; forms a diode since electrons can only flow one way; from the excess n-side to the deficit p-side; under an applied voltage. Creating light as these electrons flow to lower energy states turns our diode into a light emitting diode or LED. Adding another n-layer turns this diode into an n–p–n two-state transistor 'sandwich'. This invention led to the development of all of today's solid-state electronic devices. It is possible to go further than this and blend together quantum theory and classical gravitation. Rather than plod through this I thought it might be more fun to consider a state-of-the-art quantum sensor for the detection of a classical gravitational field. This could form the basis of the next generation of location devices.

We now discuss a highly innovative technology program led by the University of Aberdeen in the UK as part of a joint Anglo-French initiative with funding from the UK Government. This research attempts to exploit the time-domain superposition of the momentum eigenstates of a perturbed cold atom cloud as the basis of a gravity (equivalently uniform acceleration) sensor. It is part of the Quantum 2.0 revolution in manufacturing taking shape now in the UK [6], developing technology based on entanglement and superposition of quantum states.

The focus here is on the application of the system in an inertial navigation mode, as a more accurate replacement for current systems. Sensing the local acceleration of a vehicle is equivalent to sensing a local gravitational field (Einstein's 'principle of equivalence' [2]) thus we need to build a gravity sensor.

At very low temperatures (equivalently very low kinetic energies), a cloud of bosons will tend to share the same low energy state and can condense into a Bose–Einstein condensate (BEC) where all of the atoms share the same lowest energy state (the *ground state* of the system). In this extreme case the atomic cloud can be described by the single wave function Ψ corresponding to this shared quantum state. Under these conditions Ψ satisfies the Gross–Pitaevskii dynamical equation (a particular form of the Schrödinger equation).

For our time-dependent gravity sensor we assume that most of the cloud shares such a common low energy ground state (but is not necessarily in a full BEC). We also make the simplifying assumption that there is a single excited state to which atoms in the cloud can move given sufficient additional energy.

In the Schrödinger picture we have the general eigenstate equation;

$$\hat{H}\psi(z, t) = \left(\frac{-\hbar^2}{2m}\left(\frac{\partial}{\partial z}\right)^2 + V(z)\right)\psi(z, t) = E\psi(z, t)$$

for a potential function $V(z)$ and energy eigenstate ψ.

For a free particle in one-dimensional motion, this equation has the solutions;

$$\varphi(z, \hat{H}) = C_1 \exp(ikz) + C_2 \exp(-ikz)$$

corresponding to motion along $\pm z$.

We choose the $+z$ solution as our ground state.

We now excite the ground state by applying a standing wave sinusoidal energy potential. This is done in such a way that only a few atoms leave the ground state for the excited state (i.e. the perturbation is 'small'). The additional Hamiltonian during the time $\delta t = [t^-, t^+]$ when the excitation is switched on is given by

$$\hat{H}(z, \delta t) = \sin qz$$

with $qz = (k_1 - k_2)z$ corresponding to a standing wave energy potential (with q constant) being generated by two, suitably chosen, interfering momentum wavefunctions travelling in the opposite directions $\pm z$. We need to assess the effect of this additional quantum Hamiltonian.

Let

$$\theta = qz; f(\theta) = \exp\left(\frac{-i\delta t \sin\theta}{\hbar}\right) \in L^2(C, \mu),$$

the Hilbert space of all Lebesgue square integrable wavefunctions of one complex variable. The wave function solution at time $t^+ = t^- + \delta t$ is then given near to time zero by;

$$\varphi(z, 0^+) = \exp\left(\frac{-i\hat{H}\delta t}{\hbar}\right)\varphi(z, 0^-) = f(\theta)\varphi(z, 0^-)$$

$f(\theta)$ is a periodic function of θ. It can thus can be expanded as a Fourier series;

$$f(\theta) = \sum_{-\infty}^{+\infty} c_n \exp(in\theta)$$

with

$$\begin{aligned} c_n &= \frac{1}{2\pi}\int_{-\pi}^{+\pi} f(\theta)\exp(-in\theta)d\theta = \frac{1}{2\pi}\int_{-\pi}^{+\pi} \exp\left(\frac{-i\delta t \sin\theta}{\hbar}\right)\exp(-in\theta)d\theta \\ &= \frac{1}{2\pi}\int_{-\pi}^{+\pi} \exp i\left(-\left(\frac{\delta t}{\hbar}\right)\sin\theta - n\theta\right)d\theta \\ &= J_n\left(\frac{\delta t}{\hbar}\right), \end{aligned}$$

the nth Bessel function with cylinder symmetry, evaluated at $\left(\frac{\delta t}{\hbar}\right)$.

Thus

$$
\begin{aligned}
f(\theta) &= \sum_{-\infty}^{+\infty} J_n\left(\frac{\delta t}{\hbar}\right)\exp(in\theta) \\
\varphi(z, 0^+) &= f(\theta)\varphi(z, 0^-) \\
&= \sum_{-\infty}^{+\infty} J_n\left(\frac{\delta t}{\hbar}\right) C_1 \exp(in\theta)\exp ikz \\
&= \sum_{-\infty}^{+\infty} J_n\left(\frac{\delta t}{\hbar}\right) C_1 \exp i(kz + nqz) \text{ since } \theta = qz.
\end{aligned}
$$

In other words, the sine function perturbation of the initial Hamiltonian has split the original single momentum eigenstate into a *many paths* quantum superposition of momentum eigenstates separated by integer multiples of q and weighted by a constant multiple of nth order Bessel functions.

Consider now the evolution of this wave function at a significant macroscopic time T beyond 0^+. At time 0^+ the first standing wave perturbation (SW1) of the Hamiltonian is switched off and the wave function then develops as a series of momentum eigenstates in quantum superposition of the form;

$$\exp i(kz + nqz) \text{ for } n = \pm 1, \pm 2, \pm 3\ldots$$

We then apply another short standing wave perturbation (SW2) at time T. This splits each eigenstate into a further set of eigenstates. Due to the geometry of the process, some of these eigenstates superimpose at the later times $2T$, $3T$, $4T$..., forming a 'diffraction grating' in momentum space at these times. We thus organize the experimental apparatus so that we measure the 'diffraction grating' intensity of the wave function at times $t = 2T$, $3T$ and, $4T$.

These later times $2T$, $3T$ and $4T$ are known as *echo times*; the intensity of the echoes fading over time. To achieve strong readings at times $2T$, $3T$ and $4T$ requires tight control of the cold atom cloud. A cloud of cold atoms forms an approximation to an ideal gas with a distribution of momenta. This will produce a spread of echo intensity measurements.

Assume now that the vehicle is accelerating uniformly between time 0 and time $2T$ (or equivalently is embedded in a gravitational field of constant intensity). Consider the situation in the vicinity of the first echo time $t = 2T + dt$, with $|dt| \ll T$. A uniform acceleration a_z over the time interval $[0, t = 2T + dt]$ in the z-component of the vehicle body corresponds to a displacement $z = z_0 + dz$ where z_0 is the z-displacement with constant momentum and $dz = ½\, a_z t^2 = 2a_z T^2 + 2a_z T dt$ (to first order).

Ignoring the constant term C_1, it then follows from the Bessel function identity $J_{-1}(t) = -J_1(t)$ that the quantum state around the echo time $2T$ is given to first order by;

$$\begin{aligned}\psi(z, t) &= \sum_{n=\pm 1} J_n(t)\exp i(kz + nqz) \\ &= J_1(t)\exp i(kz + qz) - J_1(t)\exp i(kz - qz) \\ &= 2iJ_1(t)\exp i(kz)\sin qz\end{aligned}$$

and therefore, the grating intensity function is

$$\psi^*(z, t)\psi(z, t) = 4J_1^2(t)\sin^2 q(z) = 4J_1^2(t)\sin^2 q(z_0 + dz)$$

with a phase shift of $\delta\varphi_z = qdz \approx qa_z(2T^2)$. This phase shift can be read out through a final scattered laser pulse.

Likewise, the phase of each quantum state around other echo times $3T$ and $4T$ … will be shifted due to the same acceleration a_z. Comparing the phases of these quantum states at different echo times thus allows us to estimate the acceleration a_z with high accuracy.

Carrying out the same process in each of the other dimensions gives then a complete measurement of the uniform acceleration (a_x, a_y, a_z) or equivalent gravitational field for the vehicle body. To give a precise measurement at some time T' beyond the first few echo times requires a resetting of the cold atom cloud followed by a new cycle of perturbations.

From this initial analysis we can see that the accuracy of the acceleration measurements is driven by two key aspects. Firstly, there is the ability to tightly control the dynamics of the cold atom cloud and thus produce good readings for the initial and subsequent echo times. Secondly there is the need to quickly reset the cloud for the next set of perturbations.

There is an additional effect which has to be taken into account, namely the detected 'tidal' acceleration due to a distribution of external gravitational masses. The effect of this will be to produce an additional change of phase of the momentum eigenstates at a given echo time $2T$, $3T$ or $4T$. However, this change varies linearly across the diffraction front, and can be separated out from the effect of the internal body dynamics.

1.4 Special relativity

What is light? The answer has led to several important shifts in our understanding of how the Universe works. The first of these occurred during the 19th Century, when the experimental work of Faraday and the theoretical work of Maxwell provided an elegant framework within which to calculate electrical effects; replacing Newton's action at a distance by local fields capable of sustaining waves. This led to significant technological feats including constructing a communications link between the USA and the UK by means of an undersea cable. An astounding consequence of these results was that light propagates as an electromagnetic wave in empty space, and that similar waves, which we now call radio waves, must exist at shorter or longer wavelengths.

Early in the 20th Century, new experimental data, such as the photo-electric effect, in which light striking a surface can release electrons from the surface forming an electrical current, began to challenge this view. Einstein's 1905 analysis of the photo-electric effect led to the concept of the photon as a particle which can deliver the required energy 'punch' to release such electrons.

1.4.1 Relativistic waves

Maxwell's equations for the propagation of light waves through the vacuum of empty space can be written in the compact form;

$$\nabla\cdot E = 0; \quad \nabla\cdot H = 0$$

$$\nabla \times E = -\frac{1}{c}\frac{\partial H}{\partial t}; \quad \nabla \times H = \frac{1}{c}\frac{\partial E}{\partial t}$$

where ∇ is the gradient operator; E and H are the electric and magnetic field strength vectors and the charge density is assumed to be zero.

With $\cdot$ and $\times$ the usual dot and cross product we have the operator equivalence;

$$\nabla\cdot = \text{div and } \nabla \times = \text{curl}.$$

This symmetry between E and H leads to the plane wave solutions;

$$\nabla^2 E \equiv \frac{\partial^2 E}{\partial x^2} + \frac{\partial^2 E}{\partial y^2} + \frac{\partial^2 E}{\partial z^2} = \frac{1}{c^2}\frac{\partial^2 E}{\partial t^2}; \quad \nabla^2 H = \frac{1}{c^2}\frac{\partial^2 H}{\partial t^2}.$$

If we consider the special case of a plane wave propagating along the x-axis, with the E vector pointing in the y-direction, we have;

$$E = (E_x, E_y, E_z) \text{ with } E_x = E_z = 0 \text{ and}$$

$$E_{y(t)} = E_{y(0)} \exp 2\pi i(\nu t - kx) = E_{y(0)} \exp 2\pi i\nu\left(t - \frac{x}{c}\right).$$

Maxwell's equations imply that;

$$\frac{\partial H}{\partial t} = -c\nabla \times E \text{ thus;}$$

$$\frac{\partial H_x}{\partial t} = -c(\nabla_y E_z - \nabla_z E_y)$$

$$\frac{\partial H_y}{\partial t} = -c(\nabla_z E_x - \nabla_x E_z)$$

$$\frac{\partial H_z}{\partial t} = -c(\nabla_x E_y - \nabla_y E_x).$$

Most of these expressions disappear, leaving just;

$$\frac{\partial H_z}{\partial t} = -c\nabla_x E_y = c\frac{2\pi i\nu}{c}E_{y(t)} = 2\pi i\nu E_{y(0)} \exp 2\pi i\nu\left(t - \frac{x}{c}\right).$$

Thus the H vector has a non-zero z-component which varies with the same phase as the non-zero y-component of the E vector.

We now consider such a plane periodic wave disturbance $F(x(t), y(t), z(t), t)$ propagating through free space in the direction given by the vector $\boldsymbol{r}$ with direction cosines (l, m, n). This means that $\frac{r}{|r|}$ is the three-dimensional vector (le_x, me_y, ne_z) where e_x, e_y and e_z are unit vectors along the three space axes and $l^2 + m^2 + n^2 = 1$.

In three dimensions the set $\{x, y, z; lx + my + nz = |r|\}$ defines a plane orthogonal to the direction of propagation $\boldsymbol{r}$. The vector $\boldsymbol{k}$ defined by the components $\frac{\nu}{c}(l, m, n) = \frac{1}{\lambda}(l, m, n)$ is called the *wavevector*. For an arbitrary vector $\boldsymbol{p}$ drawn from the origin and ending in the point (x,y,z) of the plane we now have that $\boldsymbol{k} \cdot \boldsymbol{p} = \frac{\nu}{c}(lx + my + nz)$. The equation of the disturbance can thus be written as;

$$F(p(t), t) = \exp 2\pi i(\boldsymbol{k} \cdot \boldsymbol{p} \pm \nu t)$$

where the wavevector $\boldsymbol{k}$ points in the direction of propagation.

We now assume, as a convention, that our wave is of fixed wavelength (i.e. it is monochromatic light) and propagates in the positive z-direction. This means that the E and H vectors are confined to the plane defined by the x and y axes, normal to the propagation direction.

A classical light wave which is polarized in the x-direction has a space–time-dependent electric field oscillating in the x-direction which is the real part of the expression;

$$E = E_0 e_x \exp 2\pi i(kz - \nu t).$$

In the same way, I can consider a y-polarised light wave which is the real part of;

$$E = E_0 e_y \exp 2\pi i(kz - \nu t).$$

Circular polarization (i.e. circular rotation of the E vector in the $x - y$ plane) is produced by linearly combining an x-polarised plane E wave with a y-polarised plane E wave 90 degrees out of phase. The equation for such a wave is thus of the form;

$$E = E_0\left(\frac{1}{\sqrt{2}}e_x \exp 2\pi i(kz - \nu t) + \frac{i}{\sqrt{2}}e_y \exp 2\pi i(kz - \nu t)\right).$$

1.4.2 Jones vectors to describe classical polarization states

These various classical linear combinations can more compactly be described by *Jones vectors* which record the weights of the x and y components of E.

An x-direction polarized beam with weight $E_0 e_x$ thus corresponds to the Jones vector $\begin{pmatrix} E_0 e_x \\ 0 \end{pmatrix} = E_0 e_x \begin{pmatrix} 1 \\ 0 \end{pmatrix}$ while a y-direction polarized beam corresponds to $\begin{pmatrix} 0 \\ E_0 e_y \end{pmatrix} = E_0 e_y \begin{pmatrix} 0 \\ 1 \end{pmatrix}$ and the circularly polarized beam above corresponds to the Jones vector $\frac{1}{\sqrt{2}}\begin{pmatrix} E_0 e_x \\ iE_0 e_y \end{pmatrix}$. Normalizing these so that the total weight in each case is 1 leads to the vectors $\begin{pmatrix} 1 \\ 0 \end{pmatrix}$, $\begin{pmatrix} 0 \\ 1 \end{pmatrix}$ and $\frac{1}{\sqrt{2}}\begin{pmatrix} 1 \\ i \end{pmatrix}$.

1.4.3 The relativistic quantum photon

It is now possible to use the correspondence principle to argue that a polarized beam is composed of a stream of equivalently 'polarized' photon states defined by these normalized Jones vectors in a two-dimensional vector space, giving rise to the well-known two-state qubit formulation of a photon. This is a good way of building an initial intuition but does no justice to a combined quantum mechanical and relativistic approach as now follows.

The basic quantum axioms tell us we need to somehow define the 'observable' spin matrix S and the measured spin values for light correspond then to the eigenvalues of this matrix. How do we do this?

The answer was essentially supplied by Wigner who studied the group of rotations of space–time and their representation as groups of operators on vector spaces [7]. It brings into play the fact, derived from the relativistic invariance of Maxwell's equations and confirmed by experiment that;

The speed of light is independent of the speed of the source of the light; shining a torch out of a moving car does not make the photons go faster.

In two dimensions, let the vertical axis be time t and the horizontal axis be one-dimensional space x. A positive velocity is a vector pointing from the origin of these axes into the upper right quadrant to the point $(\Delta x, \Delta t)$. Speed is distance over time (related to the vector pointing angle) so changes in the speed correspond to rotations of space–time. If c, the speed of light, is invariant then $c^2(\Delta t)^2 - (\Delta x)^2 = 0 \Longrightarrow (ic\Delta t)^2 + (\Delta x)^2$ is constant in any Euclidean coordinate system. Moving a line around and preserving its length s; where $s^2 = (ic\Delta t)^2 + (\Delta x)^2$ is the Euclidean geometry of Pythagoras' theorem applied to space–time. Thus the constancy of the speed of light is equivalent to using coordinates ict and x (a *Wick rotation of the time variable*) and assuming space–time is flat (i.e. Euclidean). That is the essence of special relativity.

In this Euclidean space–time, a change of speed is still just a rotation, so we have the coordinate transformation;

$$\begin{pmatrix} \Delta x' \\ ic\Delta t' \end{pmatrix} = \begin{pmatrix} \cos\theta & -\sin\theta \\ \sin\theta & \cos\theta \end{pmatrix} \begin{pmatrix} \Delta x \\ ic\Delta t \end{pmatrix}$$

from which we can derive;

$$\Delta x' = \cos\theta\,\Delta x - ic\sin\theta\,\Delta t = \cos\theta(\Delta x - ic\tan\theta\,\Delta t)$$

$$\Delta x' = 0 \Rightarrow \Delta x = ic\tan\theta\,\Delta t \Rightarrow \tan\theta = \frac{V}{ic} \text{ where } V = \frac{\Delta x}{\Delta t}.$$

$$1 = \cos^2\theta + \sin^2\theta \Rightarrow \frac{1}{\cos^2\theta} = 1 + \tan^2\theta = 1 - \frac{V^2}{c^2}$$

$$\Rightarrow \Delta x' = \frac{1}{\sqrt{1 - \frac{V^2}{c^2}}}\left(\Delta x - ic\left(\frac{V}{ic}\right)\Delta t\right) = \frac{1}{\sqrt{1 - \frac{V^2}{c^2}}}(\Delta x - V\Delta t).$$

From the expression above containing, $\frac{1}{\sqrt{1-\frac{v^2}{c^2}}}$, called the Lorentz contraction, *we derive the implication that the speed of light is an upper bound.*

Corresponding to the non-Wick rotated four vector (ct, x, y, z) of a point in space–time, there is the energy–momentum vector; composed of four dimensionally consistent components $\left(\frac{e}{c}, p_x, p_y, p_z\right)$. If the photon is travelling with momentum p along the y-axis, then its total energy is $e = pc$ and the energy–momentum vector is then $(p, 0, p, 0)$. This is invariant only to the *Wigner little group* of rotations in the x–z plane orthogonal to the direction of propagation (y). Rotations in this plane have the form;

$$R(\theta) = \begin{pmatrix} \cos\theta & -\sin\theta \\ \sin\theta & \cos\theta \end{pmatrix}$$

We derive the eigenvalues λ_1 and λ_2 of $R(\theta)$;

$$\begin{pmatrix} \cos\theta & -\sin\theta \\ \sin\theta & \cos\theta \end{pmatrix}\begin{pmatrix} 1 \\ i \end{pmatrix} = \lambda_1\begin{pmatrix} 1 \\ i \end{pmatrix} \Rightarrow \lambda_1 = \cos\theta - i\sin\theta \Rightarrow \lambda_1 = \exp(-i\theta)$$

and

$$\begin{pmatrix} \cos\theta & -\sin\theta \\ \sin\theta & \cos\theta \end{pmatrix}\begin{pmatrix} i \\ 1 \end{pmatrix} = \lambda_2\begin{pmatrix} i \\ 1 \end{pmatrix} \Rightarrow \lambda_2 = \cos\theta + i\sin\theta = \exp(i\theta).$$

Thus the normalized eigenstates of $R(\theta)$ are $\frac{1}{\sqrt{2}}\begin{pmatrix} 1 \\ i \end{pmatrix}$ and $\frac{1}{\sqrt{2}}\begin{pmatrix} i \\ 1 \end{pmatrix}$ with eigenvalues exp $(\pm i\theta)$.

These eigenstates correspond to the Jones vectors for circular polarization of the classical EM wave, bringing the quantum and classical approaches into line.

If we consider such a rotation $R(\theta)$, then we have, to first order, for a small rotation angle θ;

$$R(\theta) = \begin{pmatrix} \cos\theta & -\sin\theta \\ \sin\theta & \cos\theta \end{pmatrix} \cong \begin{pmatrix} 1 & -\theta \\ \theta & 1 \end{pmatrix} = \begin{pmatrix} 1 & 0 \\ 0 & 1 \end{pmatrix} - i\theta\begin{pmatrix} 0 & -i \\ i & 0 \end{pmatrix}; \text{ thus}$$

$R(\theta) = I_{2\times2} - i\theta S$ as 2×2 matrices with S the spin matrix. For a large rotation through θ as the product of a number of very small rotations, we have $R(\theta) = \left(R\left(\frac{\theta}{n}\right)\right)^n = \left(I - \frac{i\theta}{n}S\right)^n$ leading in the limit to the quantum operator equivalence $R(\theta) = \exp(-i\theta S)$. From this it is clear that the spin operator S must have normalized observed eigenvalues +1 and −1.

In summary, the zero mass of the photon gives rise to an energy–momentum vector with restricted rotation symmetry in space–time and leads to the fact that it has only two quantum spin values, +1 and −1 rather than three (+1, 0, −1). This analysis shows that space–time plays an active role in particle interactions and is not just a passive spectator. *To properly take this effect into account, we need a mathematically coherent quantum theory of space–time, as well as particle interactions. We need to evolve the stage, as well as the actors who strut upon it.*

1.5 Dirac relativistic spinor theory

We described earlier the process of measuring the energy of a quantum wave function Ψ for a free particle moving in the direction x with energy E and momentum p. We have

$$\psi(x, t) = \left\{\exp\frac{2\pi i}{h}(px - Et)\right\}$$

where h is Planck's constant. We also noted that if we differentiate this expression in terms of time t then it pulls out the value E for the energy from the right-hand side, allowing it to be measured. Similarly, the linear operator $-i\hbar\frac{\partial}{\partial x}$ has momentum eigenvalue p; we can interpret this as in terms of a general momentum vector as the expression for the x-component with similar expressions for the y and z components.

We consider a particle of mass m, as measured when at rest relative to the measurement apparatus. Let it now be moving freely with velocity $V \ll c$, the speed of light. As before we have the relativistic four vector in flat space–time, and the corresponding energy–momentum vector; composed of four dimensionally consistent components

$$\frac{H}{c} = \left(\frac{E_0}{c}, p_x, p_y, p_z\right)$$

where $E_0 = mc^2$ is the relativistically invariant rest energy. H is the total system energy (the classical Hamiltonian) for a free particle, and we also have the x, y and z components of momentum explicitly noted. The squared length of this vector is given by

$$\frac{H^2}{c^2} = \left((mc)^2 + p_x^2 + p_y^2 + p_z^2\right) \tag{1.1}$$

So far I have worked within the assumptions of classical physics. Now, as discussed in section 1.1 concerning basic notions, we have to guess the underlying quantum expression giving rise to this classical equation. From the axioms of quantum theory and with Occam's Razor in mind we guess that for a free particle, with no potential energy and with rest mass m, the energy quantum operator $\hat{H}$ and momentum quantum operators $\hat{p}_{x,}\hat{p}_{y,}\hat{p}_{z,}$ satisfy the equivalent of equation (1.1) above.

Thus we assume the quantum operator relationship

$$\hat{H}^2 = m^2c^4 + c^2\hat{p}_x^{\,2} + c^2\hat{p}_y^{\,2} + c^2\hat{p}_z^{\,2}. \tag{1.2}$$

In the representation of these operators as acting on wave function quantum states ψ I can now substitute into equation (1.2) the expressions I derived for energy and momentum as quantum operators;

$$\hat{H} = i\hbar\frac{\partial}{\partial t}; \qquad \hat{p}_x = -i\hbar\frac{\partial}{\partial x} \text{ etc.}$$

Leading to the Dirac equation;

$$
\begin{aligned}
(i\hbar)^2\frac{\partial^2\psi}{\partial t^2} &= m^2c^4\psi + c^2(-i\hbar)^2\frac{\partial^2\psi}{\partial x^2} + c^2(-i\hbar)^2\frac{\partial^2\psi}{\partial y^2} + c^2(-i\hbar)^2\frac{\partial^2\psi}{\partial z^2} \\
&\Rightarrow \frac{1}{c^2}\frac{\partial^2\psi}{\partial t^2} - \frac{\partial^2\psi}{\partial x^2} - \frac{\partial^2\psi}{\partial y^2} - \frac{\partial^2\psi}{\partial z^2} = \left(\frac{mc}{i\hbar}\right)^2\psi \\
&\Rightarrow \nabla^2\psi = \beta^2\psi
\end{aligned}
$$

where ∇^2 is the Laplacian $\left(\frac{1}{c}\frac{\partial}{\partial t}\right)^2 - \left(\frac{\partial}{\partial x}\right)^2 - \left(\frac{\partial}{\partial y}\right)^2 - \left(\frac{\partial}{\partial z}\right)^2$ and $\beta^2 = \left(\frac{mc}{i\hbar}\right)^2$.

1.5.1 The relativistic spinor

Hamilton's original formulation of quaternions extended the idea of complex numbers by defining not one but three imaginary square roots of −1. Labelled *i, j* and *k*, they satisfy the following rules $i^2 = j^2 = k^2 = -1$ and they anticommute so that for example $i \times j = -(j \times i)$. A typical quaternion number is then of the form $x = x_0 + ix_1 + jx_2 + kx_3$ where the x_j are real numbers with $x^* = x_0 - ix_1 - jx_2 - kx_3$ and $xx^* = x_0^2 + x_1^2 + x_2^2 + x_3^2$; the anticommutation properties mean that cross terms of this product cancel each other. Following Hermann Weyl's contravariant and covariant vector representation, Dirac's dazzling idea was, in essence, to define contravariant *spinor operators* γ^j $(j = 0, 1, 2, 3)$ with similar quaternion-like properties so that the *Dirac Operator*

$$
\not{\partial} = \gamma^0\frac{1}{c}\frac{\partial}{\partial t} + \gamma^1\frac{\partial}{\partial x_1} + \gamma^2\frac{\partial}{\partial x_2} + \gamma^3\frac{\partial}{\partial x_3}
$$

satisfies $\not{\partial}^2\psi = \nabla^2\psi$ the cross terms again cancelling out. This gives the desired first order differential fermionic solutions ψ of the Dirac equation $\nabla^2\psi = \beta^2\psi$ as satisfying

$$
\not{\partial}\psi = \pm\beta\psi
$$

with $\beta = \left(\frac{mc}{i\hbar}\right)$. Choosing the positive solution $\not{\partial}\psi = \beta\psi$, writing $\not{\partial}$ as the dot product $\gamma\partial$, and setting, by a simple change of scale $c = \hbar = 1$, so that $\beta = \frac{m}{i}$, yields this solution in the elegant form

$$
i\gamma\partial\psi = m\psi.
$$

This is the mathematical relationship inscribed on Paul Dirac's memorial plaque in Westminster Abbey, London, laid at the graveside of Sir Isaac Newton.

Dirac's solution to the puzzle of how to represent the quantum spinor operators γ^j was to set them up as 4×4 matrix operators. In the Weyl representation, we have

$$
\gamma^j = \begin{pmatrix} 0 & \sigma^j \\ \bar{\sigma}^j & 0 \end{pmatrix} \quad \text{for} \quad j = 0, 1, 2, 3
$$

where $\sigma^j (j = 1, 2, 3)$ are the 2×2 Pauli spin matrices with $\bar{\sigma}^j = -\sigma^j (j = 1, 2, 3)$ and $\sigma^0 = \bar{\sigma}^0 = I_{2\times2}$, the 2×2 identity matrix $\begin{pmatrix} 1 & 0 \\ 0 & 1 \end{pmatrix}$.

Given these relationships we can show by direct calculation that the matrix product $i\gamma^0\gamma^1\gamma^2\gamma^3$ equals $\begin{pmatrix} -I_{2\times2} & 0 \\ 0 & I_{2\times2} \end{pmatrix}$. Defining γ^5 to be the composite matrix operator $i\gamma^0\gamma^1\gamma^2\gamma^3$, I can write the wave function state ψ in the form $\psi = \frac{1}{2}(I_{4\times4} - \gamma^5)\psi + \frac{1}{2}(I_{4\times4} + \gamma^5)\psi$. I define these two terms as the left-handed and right-handed spinor components of the state. Thus we have;

$$\psi_L = \frac{1}{2}(I_{4\times4} - \gamma^5)\psi \quad \text{and} \quad \psi_R = \frac{1}{2}(I_{4\times4} + \gamma^5)\psi$$

so that $\psi = \psi_L + \psi_R$.

Now from earlier, we know that $(I_{4\times4} - \gamma^5) = \begin{pmatrix} I_{2\times2} & 0 \\ 0 & I_{2\times2} \end{pmatrix} - \begin{pmatrix} -I_{2\times2} & 0 \\ 0 & I_{2\times2} \end{pmatrix} = \begin{pmatrix} 2I_{2\times2} & 0 \\ 0 & 0 \end{pmatrix}$.

Similarly, $I_{4\times4} + \gamma^5 = \begin{pmatrix} 0 & 0 \\ 0 & 2I_{2\times2} \end{pmatrix}$.

Thus the four vector ψ_L has only two non-zero elements (the first two) and is of the covariant form $\psi_L = \begin{pmatrix} \psi_1 \\ \psi_2 \\ 0 \\ 0 \end{pmatrix}$. Similarly, the four vector ψ_R has only two non-zero elements (the last two) and is of the contravariant form $\psi_R = \begin{pmatrix} 0 \\ 0 \\ \bar{\chi}^{\dot{1}} \\ \bar{\chi}^{\dot{2}} \end{pmatrix}$. These then correspond to the left-handed covariant Weyl spinor $\begin{pmatrix} \psi_1 \\ \psi_2 \end{pmatrix}$ and the right-handed contravariant Weyl spinor $\begin{pmatrix} \bar{\chi}^{\dot{1}} \\ \bar{\chi}^{\dot{2}} \end{pmatrix}$. In four dimensions we have $\psi = \begin{pmatrix} \psi_L \\ \psi_R \end{pmatrix}$ which is an element of a four-dimensional Hilbert space. Inserting $\psi = \begin{pmatrix} \psi_L \\ \psi_R \end{pmatrix}$ into the Dirac equation gives rise to a split solution in which the left-handed Weyl spinor quantum state ψ_L corresponds to the spin 1/2 electron and the right-handed Weyl spinor quantum state ψ_R corresponds to the spin 1/2 *positron.* This prediction of an antimatter equivalent of the electron with the same mass as the electron but with opposite charge led the American experimental physicist Carl Anderson to detect it experimentally a few years later, in 1932.

It took Dirac several years, from 1928 to 1931, to understand the important implications of his analysis. His initial aim had been modest: to look at how a relativistic approach might explain some 'fine structure' discrepancies between measurements of energy levels of the Hydrogen atom and the predictions of the non-relativistic Schrödinger equation. These discrepancies were caused by the splitting of the spectral lines into two due to the effect of the spin up and spin down components of the electron spin. Dirac's analysis [3] certainly achieved his initial aim. In addition, he came to realize that he had opened a doorway into an exuberant garden of mathematical and physical delights.

- Firstly, his analysis showed that spin as a vector quantity was not some mysterious add-on but emerged naturally from his fusion of quantum theory and special relativity.
- Secondly, his equation predicted the existence of a particle equal in mass to the electron but with positive charge—later named the positron. We now know that there are many more spin 1/2 particles than the electron. The same analysis shows in fact that every spin 1/2 particle, including all the building blocks of the stuff we see and use every day, has such a companion. This was a complete bolt from the blue—a totally unsuspected property of matter called antimatter.
- Thirdly, Dirac's concept of a positron as being a positive 'hole' in a stack of occupied electron states of negative energy predicts that an ordinary electron of positive energy could fall down the energy levels into such a hole, releasing its energy in the form of a high energy photon (a gamma ray). Thus an electron and a positron would interact by mutual annihilation, giving rise to a burst if pure energy. It is thought that soon after the Big Bang most of the antimatter was annihilated in this way, leaving a small residue of mainly ordinary matter. Conversely, a gamma ray interaction with the vacuum state could, with sufficient energy, elevate an electron from a negative energy state to a positive state plus a hole. Such gamma rays can be created for a very brief time. The overall effect is thus the brief appearance of a 'virtual' electron and positron pair which then annihilate each other. The vacuum state is thought to be a seething mass of such virtual pair creation and annihilation events. The energy density of the vacuum state affects the value of the cosmological constant in Einstein's theory of gravity. This is currently an aspect of classical gravity theory not well tied down. We do not dally to enquire further here.

1.6 von Neumann algebras

Let $\mathcal{H}$ be a Hilbert space. The set of all bounded linear operators acting on $\mathcal{H}$ is denoted $\mathcal{B}(\mathcal{H})$. This set has the multiplicative and additive properties of a non-commutative algebraic ring with identity and is referred to as the *algebra of all bounded linear operators acting on* $\mathcal{H}$. Following Bohr, I require that the measurement assumptions concerning the quantum states are captured in the theory. I thus assume that if the quantum system under study has finite energy, then $\mathcal{H}$ is finite dimensional, and this algebra consists of matrix operators. If $\mathcal{H}$ is separable with countable dimension, then matrix operators acting on $\mathcal{H}$ have finite dimensional range spaces, and the norm closure of the set of all such operators contains the compact operators. The subalgebra of compact operators retains many of the pleasant characteristics of the finite dimensional case [4]. For example, every pure state is a vector state. More generally we can consider an ascending tower of 'nested' matrix algebras and their closure in an appropriate topology gives rise to the class of *uniformly matricial algebras*; also known as *Uniformly Hyperfinite* (*UHF*) *algebras*; see for example [4, 8].

The current focus of quantum theory has moved to topological subsets $\mathcal{M}$ of $\mathcal{B}(\mathcal{H})$ called *von Neumann algebras* after John von Neumann following his creation of the subject in a series of seminal papers [9, 10]. These formed a mathematical underpinning to parallel developments in the new quantum mechanics. The beginning of this glorious period in physics can be identified with the award of the physics Nobel jointly to Dirac and Schrödinger in 1933.

It is now possible, following [11], to treat these developments as a part of an abstract quantum algebra. This has benefits later on, particularly when we need to exploit the representation theory of these algebras in the context of the Gelfand–Naimark–Segal construction. However, this whole area of mathematics is regarded (even by mathematicians) as difficult. We are ultimately interested in building better physical theories. With this in mind we shall make a start by assuming that our *quantum algebras* are both *concrete*, and *stable for the adjoint mapping.* By concrete I mean they act on a (separable) Hilbert space $\mathcal{H}$ and are subalgebras of $\mathcal{B}(\mathcal{H})$. Let $J{:}A \to A^*$ be the mapping from the subset $\boldsymbol{S}$ to $\mathcal{B}(\mathcal{H})$ derived from taking the adjoint of an operator. $\boldsymbol{S}$ is *stable* for this mapping if $J{:}\ \boldsymbol{S} \to \boldsymbol{S}$.

On the basis of this discussion, I define a *quantum algebra* initially as a concrete subset of $\mathcal{B}(\mathcal{H})$ which forms a linear and associative algebraic ring over the field of complex numbers. A quantum algebra is an algebra which is both concrete and stable. Later we will remove the requirement for a quantum algebra to be concrete. A subalgebra $\mathbf{A}$ of $\mathcal{B}(\mathcal{H})$ which is a normed linear space for the usual operator norm $\|.\|$ on $\mathcal{B}(\mathcal{H})$ we define as a *quantum Banach Algebra* if

- $\|AB\| \leqslant \|A\|\|B\|(A, B \in \mathbf{A})$.
- $\mathbf{A}$ is complete relative to the norm topology.

A *quantum C*-algebra* is a quantum Banach algebra such that $\|A^*A\| = \|A\|^2$.

So far, I have just considered the topology on these algebras derived from the operator norm, relative to which all bounded linear operators are continuous. I denote this topology by τ_n. Any topology on a set is completely defined by an equivalent definition of convergence [12]. Thus, the family of seminorms $\{p_x;\, p_x(T) = \|Tx\| \text{ for } x \in \mathcal{H}\}$ defines the strong operator topology τ_s on $\mathcal{B}(\mathcal{H})$ and any subalgebras $\mathcal{M}$ of $\mathcal{B}(\mathcal{H})$, by defining $T_n \to T \Leftrightarrow p_x(T_n - T) = \|T_n x - Tx\| \to 0\ \forall\ x \in \mathcal{H}$. In a similar way [13], by defining different sets of seminorms, I can define a range of topologies applicable to subalgebras $\mathcal{M}$ of $\mathcal{B}(\mathcal{H})$;

- The ultrastrong topology $\tau_{\sigma s}$ is defined by the seminorms

$$\left\{p_X(T) = \sum \|Tx_j\|^2;\ \ X = \left\{x_j;\ \ \sum \|x_j\|^2 < \infty\right\}\right\}.$$

- The weak topology τ_w is defined by the seminorms $p_{x,y}(T) = |\langle Tx, y\rangle|$;
- The ultraweak topology $\tau_{\sigma w}$ is defined in a similar way to the ultrastrong, replacing the sum of norms by a sum of inner products. It is thus defined by the seminorms;

$$\left\{p_X(T) = |\sum\left\langle Tx_j, y_j\right\rangle|;\ X = \left\{x_j, y_j;\ \sum(\|x_j\|^2 + \|y_j\|^2) < \infty\right\}\right\}.$$

A (quantum) von Neumann algebra is defined as a quantum C*-subalgebra $\mathcal{M}$ of $\mathcal{B}(\mathcal{H})$ which is closed for the weak operator topology [4]. The algebra $\mathcal{M}$ is rich in projections, the largest of which forms an identity for $\mathcal{M}$.

One of the most powerful and elegant properties of von Neumann algebras is the dialogue between their algebraic and topological properties. This is illustrated by the following discussion.

If $\mathcal{S}$ is a subset of $\mathcal{B}(\mathcal{H})$, the commutant of $\mathcal{S}$, denoted S', is defined as the set of all elements T of $\mathcal{B}(\mathcal{H})$ such that $[T, S] = TS-ST = 0$ for all $S \in \mathcal{S}$. This is a τ_w-closed subset of $\mathcal{B}(\mathcal{H})$. The commutant of S' which is the double commutant of $\mathcal{S}$, is a purely algebraic construct, yet it can be shown that S'' is the smallest von Neumann algebra containing $\mathcal{S}$. This is known as von Neumann's *Double Commutant Theorem*. It follows that if $\mathcal{S}$ is a C*-algebra then the von Neumann algebra S'' is the closure of $\mathcal{S}$ for the τ_w-topology. The exact way that the algebraic double commutant equates to the weak closure is discussed by G Pedersen, leading to his *Up-Down Theorem.*

1.7 A higher level of abstraction: quantum W*-algebras

The history of the development of quantum theory is driven by the need for more abstraction in order to be closer to the algebraic structure. I can now develop the entire theory of quantum algebras as *abstract* objects called *quantum W*-algebras* [11] which we can examine, on an even-handed basis, through their various representations as *concrete* von Neumann algebras. We can see that the properties which define a quantum Banach algebra and a C*-algebra are algebraic in form and can be easily transmuted into definitions for their abstract equivalents. However, for a von Neumann algebra the transmutation into an abstract quantum W*-algebra is more complex. First, some definitions.

An observable element A of a quantum W*-algebra **A** with identity element I is defined as *positive* if the spectrum of A is a subset of the non-negative real numbers.

An element ρ of the dual space **A*** is a *quantum state* if $\rho(A) > 0$ for A a non-zero positive observable, and $\rho(I) = 1$. The set of all such states forms a compact subset of **A*** for the weak* topology $\sigma(A^*, A)$.

We define the subset of these generalised quantum states which are continuous for the $\tau_{\sigma w}$ topology to be *normal.* It turns out that the closed linear span of these normal states has the properties of a Banach subspace of **A***. This Banach subspace is called the *predual* of **A** and we denote it as $\mathbf{A}_*$ From the duality $\rho(A) \leftrightarrow \hat{A}(\rho)$ we can then identify **A** as the Banach dual space of its predual $\mathbf{A} = (\mathbf{A}_*)^*$.

We therefore define an *abstract(quantum) W*-algebra* to be an abstract C*-algebra which is also isomorphic to the dual space of a Banach space [11]. I will consider these algebras in more detail later. For now we assume the invariance of such a quantum system to views from different inertial reference frames, which is the essence of Special Relativity, as discussed earlier.

1.8 A brief comparison with the approach of Rovelli and Penrose[1]

We agree on the following.

- Relativity has been tested in the strong gravity regime caused by colliding black holes. Two separated LIGO[2] detectors have identified the resultant gravitational waves.
- The lack of evidence for supersymmetry at the LHC during recent runs is surprising given all of the theoretical arguments in its favour. It tends to undermine approaches to quantum gravity for which supersymmetry is essential from the start. Given the gulf in energy between the current LHC and the Planck regime, I assume a massless supersymmetric graviton although this is not essential to my approach.
- Loop quantum gravity is the quantum theory of discrete space and time, unifying quantum theory and general relativity.
- The unitary spin subgroup $SU(2)$ of the matrix group $SL(2, C)$ is a fundamental 'building block'.

At this point we begin to diverge

I begin by assuming that in the very early era of the Big Bang known as the Planck regime, the Universe is still not transparent. Thus the key relativistic invariants relate to spin, rather than the speed of light. These $SU(2)$ spinor elements evolve over time, increasing their stability by randomly linking together, forming a computational spin network[3]. Pathways through this network are analogous to algorithm pathways in a quantum computer. The network is dimensionally embedded by exploiting the mathematics of homology theory, creating discrete paths in an emergent discrete space–time. I give a full mathematical account of this approach in the next chapter. My approach differs from the spin networks approach

[1] Rovelli, Penrose and Ashtekar were three of the founding fathers of loop quantum gravity along with Smolin and Sen. We discuss Ashtekar's perspective in chapter 4 as a key motivation for that chapter.

[2] LIGO; Laser Interferometry Gravitational-wave Observatory is a US based experiment which detects ripples in space–time ('gravitational waves') through their lengthening or shortening effect on a linear object in their path. There are two well separated detectors which must both trigger almost simultaneously for a valid signal. International staff includes UK, Australia.

[3] We define a *computational spin network* to be a finite or countable quiver consisting of a directed graph, where the nodes represent entangled spin inputs and a directed random link between two nodes corresponds to a quantum gate.

of Penrose. As discussed in chapter 2, assuming planar isotopy, it is possible to associate various locally deformable lines in the plane with Weyl 2-spinor calculations, giving rise to topological structures called *(Penrose) spin networks*. The point of this approach is that spinor calculations become a sequence of potentially simpler topological transformations with connections to knot theory[4].

Space in Rovelli's treatment [14] is considered separately from time and is defined in terms of the proximity of one object to another. Rovelli thus begins with the dual network of proximities for a set of extended objects in space. He then exploits the mathematical properties of $SU(2)$ as a compact Lie subgroup of $SL(2,C)$ by generalizing the non-relativistic Schrödinger picture in the following way. He defines the key Hilbert space for his interaction picture as $L^2[SU(2)]$; the Hilbert space of square integrable functions ψ: $SU(2) \to C$.

He then defines a 'generalised' angular momentum operator L^x acting on $L^2[SU(2)]$ in terms of σ^x one of the (normalised) Pauli spin matrices.

Similarly he defines the operators L^y, L^z as 'components' of a total angular momentum operator L. These operators are key to his prediction that area and volume as observables have discrete eigenvalues, and thus there are finite lower bounds to area and volume measurements. If true, this implies that such representations of space have no need to consider the limits of renormalisation groups.

References

[1] Friar William of Occam (14th Century) Occam's razor; Entities should not be multiplied without necessity.

[2] Einstein A 1960 *The Meaning of Relativity* 6th edn (London: Methuen) reprinted

[3] Dirac P 1958 *The Principles of Quantum Mechanics* 4th edn (Oxford: Oxford University Press)

[4] Kadison R and Ringrose J 1983 *Fundamentals of the Theory of Operator Algebras* vols I, II (New York: Academic)

[5] Shockley W 1951 *Electrons and Holes in Semiconductors with Applications to Transistor Electronics* (New York: Bell Labs/van Nostrand)

[6] Till S and Pritchard J 2014 *The Quantum Technology Landscape*; Final Version, Dstl, Reference Dstl/PUB75620 Unclassified and Uncontrolled

[7] Wigner E 1939 On the unitary representations of the inhomogeneous Lorentz group *Ann. Math.* **40** 149

[8] Glimm J 1960 On a certain class of operator algebras *Trans. Am. Math. Soc.* **95** 318–40

[9] Murray F and von Neumann J 1936 On rings of operators I, II, IV *Ann. Math. (I, IV)* **37** 116–229

Murray F and von Neumann J 1943 On rings of operators I, II, IV *Trans. Am. Math. Soc. (II)* **44** 716–808

[10] von Neumann J 1940 On Rings of operators III *Ann. Math.* **41** 94–161

[4] Two three-dimensional knots are equivalent if there is a sequence of such two-dimensional transformations linking one knot to the other (the 'Reidemeister moves').

[11] Sakai S 1971 *C*-Algebras and W*-Algebras* (Berlin: Springer)
[12] Kelley J 1970 *General Topology* (Princeton, NJ: van Nostrand)
[13] Dunford N and Schwartz J 1957 *Linear Operators Part 1* (New York: Wiley)
[14] Rovelli C 2018 *An Introduction to Loop Quantum Gravity* A series of Lectures given at CPT Marseilles; available on YouTube™

IOP Publishing

A Mathematically Coherent Quantum Gravity

James Moffat

Chapter 2

Computational spin networks and quantum paths in space–time

If the reader wittingly reflects on all that is laid down, (s)he will discover here the end of the Universe[1].

2.1 Introduction

We begin by recalling some of the key points concerning quantum theory in chapter 1.

Quantum theory is essentially a guess at how inputs to an experiment at the quantum level relate to outputs. These inputs and outputs are real numbers shown on displays.

Good guesses are those which are consistent with all the existing experimental data (the facts) and have high predictive power, within their domain of application. Our currently accepted theory is the simplest of the good guesses, because the simplest explanation consistent with the facts is, we assume, nearest to the truth.

Finally, the more abstract the theory, the more broadly relevant it is to the real world.

Thus, a good model of quantum gravity will have the following characteristics;

- It will have an elegant mathematical structure, since this is the simplest way of expressing its logic.
- It will be a theory based on the transformation of existing information into new information.
- It will be a black box model. All we require is that for a given set of inputs, the outputs are correct (i.e. consistent with the facts), within its domain of application. An example is the non-relativistic quantum theory of a particle

[1] From the Latin script in the tessellated mosaic laid in front of the high altar of Westminster Abbey, London, for Henry III (13th Century). Translation by Foster R 1992 *Patterns of Thought* (London: Jonathan Cape).

doi:10.1088/978-0-7503-2580-6ch2

trapped in the finite potential well of an atomic nucleus. The Schrödinger wave function intensity $\psi^*\psi$ for the particle location has tails lying outside the potential well. This allows us to calculate the effect of quantum tunnelling. The theoretical prediction correctly computes the measured probability of the particle escaping but the mechanism by which this process—which we call radioactivity—is achieved is hidden from us. I refer to this as *quantum censorship*. If we open the black box the system reverts to classical mechanics.

I define a black box model to be *scale free* if the transformation from inputs to outputs is invariant under a change of scale of the variables representing these inputs and outputs. Such a system is then *renormalisation invariant.*

From the point of view of an abstract W*-algebra $\mathbf{A}$ with identity I, we can interpret quantum theory as follows;

- A quantum state ρ is a positive element of the dual space $\mathbf{A}^*$ with $\rho(I) = 1$.
- A representation π is a homomorphism from $\mathbf{A}$ to a von Neumann subalgebra of $\mathscr{B}(\mathscr{H})$ for some Hilbert space $\mathscr{H}$. The Hilbert space is either finite dimensional or separable.
- Each quantum state ρ gives rise to a representation in which it is equivalent to a Dirac ket $|x\rangle$ of the Hilbert space with norm one; this is the Gelfand–Naimark–Segal (GNS) construction [1]. An irreducible representation corresponds to ρ being a pure state. Since the state space is a weak*-convex compact subset of the dual space, it is the closed convex hull of its extreme points These extreme points are the pure states.
- A system observable, such as the total system energy, corresponds to a bounded linear operator acting on this Hilbert space. Measurements of the system energy correspond to eigenvalues of the linear operator.

2.2 The measurement of space and time

Our theory aims to unite quantum theory, as expressed in terms of a W*-algebra, and Einstein's theory of gravity, called general relativity, together into a larger theory which we call loop quantum gravity. This larger theory is the quantum theory of space–time. We have noted that such a theory is a black box model of reality. We therefore avoid the need to worry about the essential nature of space–time. Instead we just need to define the measurement of space–time in a way which is compatible with both quantum theory and relativity. We will be guided by Einstein's thoughts in the first chapter of [2], and by Dirac [3]. In a review of the nature of space and time in pre-relativity physics Einstein concludes that Euclidean geometry is the theory of invariants with respect to 'preferred systems of coordinates, the Cartesian systems which transform into each other by linear orthogonal transformations'. It follows that an equation equivalent to setting a Cartesian tensor to zero (or equivalently setting two Cartesian tensors of equal rank to equality) in Euclidean space is then independent of the coordinate system. By performing a Wick rotation $ct \to ict$ of the time variable in special relativity, as discussed in chapter 1, we can treat Minkowski space as a four-dimensional Euclidean space. Thus, systems of coordinates for space–time events in special relativity also correspond to Cartesian systems which are represented by real variables for points in space and pure imaginary variables for time. This is consistent with the unification of special relativity and quantum theory achieved by Dirac resulting in the prediction of antimatter. Such Cartesian systems correspond to measurement using unit rods and standard clocks subject to considerations of local simultaneity. Two avatar observers called Alice and Bethe bringing together their clocks and measuring rods will note that they are identical. In general relativity, space–time corresponds to a smooth manifold with generalised real Gaussian coordinates (x_j; $j = 0, 1, 2, 3$). Within a tangent plane to this manifold we can impose a set of local Cartesian coordinates (X_j;$j = 0, 1, 2, 3$). Expressing the square of the invariant infinitesimal interval *ds* of classical space–time in both sets of coordinate systems yields;

$$ds^2 = \eta_{\mu\mu} dX_\mu^2 = \eta_{\mu\nu} dx_\mu dx_\nu$$

(Einstein summation convention applies). This corresponds to expressing each of the real variables dx_μ as a linear combination of the real variables dX_μ,dX_ν (or vice versa; the mapping is invertible).

At the level of human perception, time separates from space and corresponds to the ordering of events by increasing entropy; the whole egg always precedes the egg smashing and making a mess on the floor. Since the Universe is, by definition, a closed system, ultimately entropy will increase, although locally open subsystems allowing the flow of information and energy across the system boundary might lead to a locally low entropy sponge cake rather than a high entropy mess. But at the quantum level, working with Feynman diagrams of particle interactions, the relativistic idea of very small increments of time as part of a very small space–time vector *ds* joining neighboring events sounds like something we can work with.

Einstein's non-quantum theory of gravity (general relativity) is centred around the principle of equivalence; that locally a constant gravitational field is equivalent to uniform inertial acceleration. This is applied to replacing the ring laser gyro and similar devices in chapter 1 by coupling a quantum 'many paths' approach in momentum space to the principle of equivalence. General relativity can be criticized as being not consistent with the principle of parsimony in our black box approach. Like a less favoured Aunt it brings too much baggage. A sparser quantum-based theory, equivalent to waymarking a network of possible paths, will be introduced based on the idea of small finite increments—a *granularity*—of space–time.

We will, in fact, show in this chapter that the effect of quantum assumptions is to impose a granularity on space–time at the Planck scale corresponding to a lower limit to measurement and a finite theory in the sense of Dirac. A Wilsonian renormalisation flow which is based on pushing nearby points of space–time together thus has a lower limit corresponding to a negative gravitational 'pressure' at the Planck scale. This negative gravitational pressure may be the driver of inflation in the early Universe.

We take from relativity both the fusion of space and time into a single entity and the representation of gravity as a set of oriented, locally bounded, compact double-cones. However, the discrete nature of our computational spin network X implies that only a countable number of these double-cones are required; each one attached to an embedded node of X. These act like waymarks, directing the local discrete movements through the embedded network *embdX* from node to node by the action of the translation subgroup of the Poincaré group. From this perspective, flat space–time is a very special case in which each of the local double-cones has exactly the same orientation as its neighbours. Thus, the existence of a gravitational field is the generic case. This is similar to one of the key ideas behind special relativity[2]; that a moving frame of reference is the generic case, not requiring further justification.

In chapter 1, we considered the dynamics of the quantum state of a spin 1/2 fermion in a way which is consistent with the requirements of special relativity. This quantum state is four-dimensional, with the first two components corresponding to the particle (e.g. the electron) and the second two components corresponding to the anti-particle (e.g. the positron). We assume the Weyl representation of the Dirac 4-spinors of this theory applies so that;

$$\gamma^{\mu} = \begin{pmatrix} 0 & \sigma^{\mu} \\ \bar{\sigma}^{\mu} & 0 \end{pmatrix}$$

where $\sigma^{\mu}(\mu = 0, 1, 2, 3)$ are the extended 2×2 Pauli $2-$ spinors, $\bar{\sigma}^0 = \sigma^0 = I_{2\times 2}$ and $\bar{\sigma}^{\mu} = -\sigma^{\mu}(\mu = 1, 2, 3)$.

The representations $M \to M$ and $M \to (M^{-1})^T$ of the special linear group of 2×2 matrices $SL(2, C)$ have the same dimensionality and are therefore unitarily equivalent; there is a unitary matrix $\varepsilon = \varepsilon_{AB}$ such that for all M, $\varepsilon^{=1} M \varepsilon = (M^{-1})^T$ where $\varepsilon = \begin{pmatrix} 0 & -1 \\ 1 & 0 \end{pmatrix}$. Defining the contravariant tensor ε^{AB} by $(\varepsilon^{AB})^{-1} = \begin{pmatrix} 0 & -1 \\ 1 & 0 \end{pmatrix}$ we have $\varepsilon^{AB} = \begin{pmatrix} 0 & 1 \\ -1 & 0 \end{pmatrix}$.

The matrices ε_{AB} and ε^{AB} are covariant and contravariant tensor 2-forms in the Weyl representation which can be used to raise and lower indices in the usual vector and tensor calculus, within which left-handed chiral 2-spinors are covariant vectors, and right-handed chiral 2-spinors are contravariant vectors in the notation introduced by H Weyl. The following calculation illustrates this 'spinor calculus' for $\psi \in F_*$ and $\chi \in F$;

$$\psi\chi = \psi^A \chi_A \;= \varepsilon^{AB} \psi_B \chi_A = \varepsilon^{12} \psi_2 \chi_1 + \varepsilon^{21} \psi_1 \chi_2 = \psi_2 \chi_1 - \psi_1 \chi_2 .$$

We also define the matrix $\bar{\varepsilon}$ such that the representations $M \to M_*$ and $M \to (M_*^{-1})^T$ are equivalent so that $\bar{\varepsilon}^{-1} M_* \bar{\varepsilon} = (M_*^{-1})^T$ for all M_*.

Assuming planar isotopy, it is possible to associate various locally deformable lines in the plane with Weyl 2-spinor calculations, giving rise to topological structures called *spin networks*. As a simple example, the tensors $\bar{\varepsilon} = \varepsilon^{\dot{A}\dot{B}} = \begin{pmatrix} 0 & 1 \\ -1 & 0 \end{pmatrix}$ and $\bar{\varepsilon}^{-1} = \varepsilon_{\dot{A}\dot{B}} = \begin{pmatrix} 0 & -1 \\ 1 & 0 \end{pmatrix}$ correspond to;

[2] The other being the invariance of the speed of light.

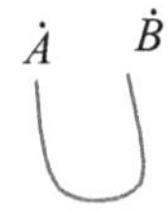

and

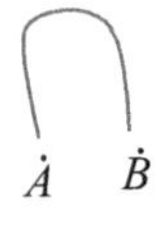

Their product is then a closed loop which corresponds to a scalar λ times the identity matrix with $\lambda = 1$ in this case. The advantage of this approach is that spinor calculations become a sequence of potentially simpler topological transformations with connections to knot theory[3].

2.3 Computational spin networks

We build on these ideas, together with those of T Regge on tessellation approximations to four-dimensional manifolds in general relativity [4] which we turn on their head to consider general relativity as an approximation to a quantum tessellated space. We define a *computational spin network* to be a finite or countable quiver consisting of a directed graph, where the nodes represent entangled spin inputs and a directed random link between two nodes corresponds to a quantum gate, as we now discuss. We will assume in what follows that the number of nodes n is finite. There is no difficulty in taking the thermodynamic limit $n \rightarrow \infty$ where necessary.

If we represent the basis spin $\pm 1/2$ eigenvectors as the column vectors $0> = \begin{pmatrix} 1 \\ 0 \end{pmatrix}$ and $1> = \begin{pmatrix} 0 \\ 1 \end{pmatrix}$, then the NOT quantum gate which switches $|0\rangle$ to $|1\rangle$ and $|1\rangle$ to $|0\rangle$ corresponds to the unitary matrix $U = \begin{pmatrix} 0 & 1 \\ 1 & 0 \end{pmatrix}$. It is easy to check that $U|0\rangle = |1\rangle$ and $U|1\rangle = |0\rangle$. The Hadamard quantum gate rotates each spin eigenvector into a quantum superposition and corresponds to the unitary matrix $\frac{1}{\sqrt{2}}\begin{pmatrix} 1 & 1 \\ 1 & -1 \end{pmatrix}$. All quantum gates correspond in this way to multiplying the input state vector by a unitary matrix mapping, usually of the form $SU(2)$.

To ease notational clutter, in all that follows we denote the joint tensor product state $|x\rangle \otimes |y\rangle$ as $|x\ y\rangle$.

Given the entangled Bell state $|\beta(0, 0)\rangle = \frac{1}{\sqrt{2}}(|00\rangle + |11\rangle)$ and $|y_0\rangle = \frac{1}{\sqrt{2}}(\alpha|0\rangle + \beta|1\rangle)$

we perform a set of steps from one node to the next neighboring node through the network as follows;

Step 1. Let $|y_1 = |y_0\rangle \otimes |\beta(0, 0)\rangle = \frac{1}{2}(\alpha\,|0\rangle \otimes |\beta(0, 0)\rangle + \beta\,|1\rangle \otimes |\beta(0, 0)\rangle)$

$$= \frac{1}{2}(\alpha\,|0\rangle \otimes (|00\rangle + |11\rangle)) + (\beta\,|1\rangle \otimes |00\rangle + |11\rangle)$$

[3] Two three-dimensional knots are equivalent if there is a sequence of such two-dimensional transformations linking one knot to the other (the 'Reidemeister moves').

$$\equiv\frac{1}{2}(\alpha \mid 000\rangle + \alpha \mid 011\rangle) + (\beta \mid 100\rangle + \beta \mid 111\rangle).$$

Step 2. Let $|y_2>= U(CNOT)|y_1>$ where the mapping $U(CNOT)$ is defined by;

$$U(CNOT)|\,00\delta\rangle \to |\,00\delta\rangle;\ |\,01\delta\rangle \to |\,01\delta\rangle;$$

$$|10\delta\rangle \to |11\delta\rangle \text{ and } |11\delta\rangle \to |10\delta\rangle \text{ (for } \delta = 0 \text{ or } 1);$$

Thus;

$$|\,y_2\rangle = \frac{1}{2}(\alpha \mid 000\rangle + \alpha \mid 011\rangle) + (\beta \mid 110\rangle + \beta \mid 101\rangle).$$

We now add an additional *Step 0* at the beginning by applying the Hadamard unitary gate $U(H)$ to the superposition $|\,y_0\rangle$.

We now have the sequence of network steps $y_0 \to y_1 \to y_2 \to y_3$ as follows;

$|y_3\rangle = U(CNOT)|y_2\rangle$; $|y_2\rangle = |y_1\rangle \otimes |\beta(0, 0)$; $|y_1\rangle = U(H)|y_0\rangle$ and $|y_0\rangle = \frac{1}{\sqrt{2}}(\alpha|0\rangle + \beta|1\rangle)$.

We have $|y_1\rangle = U(H)|y_0\rangle = \frac{1}{\sqrt{2}}(\alpha U(H)|0\rangle + \beta U(H)|1\rangle)$

$$|y_2\rangle = \frac{1}{2\sqrt{2}}\{\alpha U(H)|0\rangle(|00\rangle + |11\rangle) + \beta U(H)|1\rangle(|10\rangle + |01\rangle)\}$$

$$=\frac{1}{2\sqrt{2}}\{(\alpha|0\rangle + \alpha|1\rangle)(|00\rangle + |11\rangle) + (\beta|0\rangle - \beta|1\rangle)(|10\rangle + |01\rangle)\}$$

$$=\frac{1}{2\sqrt{2}}\Big\{\alpha|000\rangle + \alpha|011\rangle + \alpha|100\rangle + \alpha|111\rangle + \beta|010\rangle + \beta|001\rangle - \beta|110\rangle - \beta|101\rangle\}$$

$$|y_3\rangle = U(CNOT)|y_2\rangle$$

$$=\frac{1}{2\sqrt{2}}\Big\{\alpha|000\rangle + \alpha|011\rangle + \alpha|110\rangle + \alpha|101\rangle + \beta|010\rangle + \beta|001\rangle - \beta|100\rangle - \beta|111\rangle\}$$

$$=\frac{1}{2\sqrt{2}}\left\{\begin{Bmatrix}\alpha|0\rangle|00\rangle + \alpha|0\rangle|11\rangle + \alpha|1\rangle|10\rangle + \alpha|1\rangle|01\rangle + \beta|0\rangle|10\rangle \\ + \beta|0\rangle|01\rangle - \beta|1\rangle|00\rangle - \beta|1\rangle|11\rangle\end{Bmatrix}\right\}.$$

Collecting terms this yields;

$$|y_3\rangle = \frac{1}{2}\{|00\rangle U(00)|y_0\rangle + |01\rangle U(01)|y_0\rangle + |10\rangle U(10)|y_0\rangle + |11\rangle U(11)|y_0\rangle\}$$

with

$$U(00) = I_{2\times2};\ U(01) = \begin{pmatrix} 0 & 1 \\ 1 & 0 \end{pmatrix}^{-1};$$
$$U(10) = \begin{pmatrix} 1 & 0 \\ 0 & -1 \end{pmatrix}^{-1};\ U(11) = \begin{pmatrix} 0 & 1 \\ -1 & 0 \end{pmatrix}^{-1}.$$

These are all real unitary matrices, thus their inverse is simply the transpose matrix in each case;

$$U(00) = I_{2\times2} = \begin{pmatrix} 1 & 0 \\ 0 & 1 \end{pmatrix};\ U(01) = \begin{pmatrix} 0 & 1 \\ 1 & 0 \end{pmatrix};\ U(10) = \begin{pmatrix} 1 & 0 \\ 0 & -1 \end{pmatrix};\ U(11) = \begin{pmatrix} 0 & -1 \\ 1 & 0 \end{pmatrix}.$$

Note that $U(00) = \sigma^0\, U(01) = \sigma^1\, U(10) = \sigma^3\, iU(11) = \sigma^2$ where $\{\sigma^j;\, j = 0, 1, 2, 3\}$ are the extended Pauli spin matrices introduced in chapter 1 with $\sigma^0 = I_{2\times2}$.

We can interpret this process in physics terms as follows. We have constructed the sequence of Dirac quantum states $|y_0\rangle \to |y_1\rangle \to |y_2\rangle \to |y_3\rangle$ as steps across our computational spin network. Each of the two superpositions gives rise to a fork in the path thus there are four superposed options. For each of these options, we need to show that the initial state $|y_0\rangle$ is retrievable from the superimposed end state $|y_3\rangle$. For this last part we ask our avatar scientists Alice and Bethe to aid the discussion.

We have;

$$|y_3\rangle = \frac{1}{2}\{|00\rangle U(00)|y_0\rangle + |01\rangle U(01)|y_0\rangle + |10\rangle U(10)|y_0\rangle + |11\rangle U(11)|y_0\rangle\}.$$

Assume that Alice prepared both the network and the initial quantum state $|y_0\rangle$ and wishes to send this state to Bethe. She constructs the pathway $|y_0\rangle \to |y_1\rangle \to |y_2\rangle \to |y_3\rangle$ ending at Bethe. She then performs a measurement across the product eigenstates $|00\rangle;|10\rangle;|01\rangle$ and $|11\rangle$. She texts the eigenvalue result to Bethe as '00', '10', '01' or '11'. Bethe then applies the appropriate 2×2 unitary matrix to the state supplied by the network to turn it into the state $|y_0\rangle$. For example, if the text message '01' is sent, she applies the matrix $U(01)^{-1} = U(01) = \sigma^1$, one of the Pauli spin matrices. This process is known as *quantum teleportation.*

2.4 The homology invariants of space–time

We now prove that this transport from node to node of the computational spin network is equivalent to the action of a sequence of discrete translations in space–

time from one node to the next neighboring node. A discrete path in space–time can then be considered as a series of applications of the translation subgroup of the Poincaré group. This then forms the stage which allows the 'players' (the stage and the particles of the theory) to interact [5–10]. In chapter 1, we discussed an example of Wigner's little group which illustrates the way in which space–time can directly influence particle properties such as the spin eigenstates.

Our mathematical approach is to consider the homology invariants of the network structure. Homology is fundamentally concerned with the classification of topological spaces through decomposition of the space into geometric objects such as triangles, which tesselate the surface of the structure. In loop quantum gravity this is closely related to Rovelli's idea, discussed in chapter 1, of an area operator whose eigenvalues correspond to the discrete measurement of area [11]. For these reasons we choose to focus on the homological invariants of a computational spin network. From this analysis we will show that the 'machinery' of such algebraic approaches is fully available to us at this fundamental network level. It allows us to develop the desired mathematical linkage between transport across nodes of this spin network and discrete transitions in spacetime. The contravariant dual to a homology is a cohomology, and the machinery of cohomology has been applied to the task of considering the structure of multilinear mappings from a W*-algebra, as defined in chapter 1, into a Banach module; with derivations as the basis for the coboundary mapping [12]. I conjecture a connection to the ideas presented here. In similar vein, the Gromov–Witten invariants of string theory [13–15] exploit the machinery of homology theory in order to help provide a classification of the landscape of Calabi–Yau topological manifolds.

So, let us begin by considering a computational spin network $\boldsymbol{X}$ defined as a finite quiver consisting of a directed graph with a set of nodes n corresponding to entangled spin inputs and directed links between these nodes representing quantum computational gates. As a representation of space–time this directed graph cannot represent time-travel into the past, thus some potential pathways are denied in order to avoid illogical consequences. Following a suggestion of references [16, 17] we assume that the network topology is that of an evolving random network, of class (n,pr) derived from a fixed probability pr of a link between two randomly chosen nodes, with an upper bound of $\frac{n(n-1)}{2}$ links. Percolation theory [18] applied to this structure implies a critical value of pr denoted p_c for which the network becomes path connected. The value $pr = 1$ corresponds to a fully connected graph.

In the normal way we consider this graph as the skeleton of a simplicial tessellated surface when embedded in a dimensioned topological space. For $p_c < pr \leqslant 1$ we have a range of units of tessellation including 3-simplexes (tetrahedra) and potentially higher dimensional generalisations. We can use these basic properties to develop the structures, including simplicial complexes, of an Eilenberg singular homology [19, 20] associated with a given computational spin network. In what follows, we cannot in principle resolve the embedding space into subunits of space–time smaller than triangles (2-simplexes) with finite one-dimensional edges. We refer to such a topological space as a *discrete embedding space*. This is consistent with the discussion of special relativity in chapter 1 in which increments of space and time can be traded off against each other as part of a more fundamental unit; the space–time vector forming a one-dimensional boundary of a triangular 2-simplex.

Given a dimensioned embedding of X, we assume that the resultant topological space *embdX* is both finite dimensional and discrete in the sense discussed above. For the set of p-simplexes Δ_i^p forming part of the structure of our abstract computational network, we can then define a set of embedding functions $\{e_i^p : \Delta_i^p \to embdX; p \geqslant 2\}$. The p-simplex Δ_i^p has $p + 1$ boundary faces forming (p–1)-simplexes; each of these having a lift function $f_{ij}^{p-1} : \Delta_{ij}^{p-1} \to \Delta_i^p$ and corresponding embedding function;

$$e_i^p \circ f_{ij}^{p-1} : \Delta_{ij}^{p-1} \to embdX.$$

In this context, space–time is the unique geometric embedding of the singular homology induced by the computational spin network. We define the boundary function ∂_p for $p \geqslant 2$, acting on the embedded p-simplex Δ_i^p in terms of its $p + 1$ boundary faces as the alternating sum of the boundary face embedding functions;

$$\partial_p \circ e_i^p = \sum_{j=0}^{j=p} (-1)^j e_i^p \circ f_{ij}^{p-1}.$$

The alternating sum definition ensures that the boundary of a boundary is trivial;

$$\partial_p \circ \partial_{p+1} = 0 \;\; \forall \;\; p \geqslant 1.$$

Define now the freely generated abelian group C_p on the set of all p-embeddings e_i^p. A typical element of C_p is a sum of elements of the form $n_i e_i^p$ with n_i a positive integer. If $\xi \in C_p$ then we have;

$$\partial_p \circ \xi = \partial_p \circ \sum_i n_i e_i^p = \sum_i n_i \partial_p \circ e_i^p = \sum_i n_i \sum_{j=0}^{j=p} (-1)^j e_i^p \circ f_{ij}^{p-1}.$$

Thus ∂_p maps $C_p \to C_{p-1}$, allowing us to create the chain complex;

$$0 \xleftarrow{\partial_0} C_0 \xleftarrow{\partial_1} C_1 \ldots C_{p-1} \xleftarrow{\partial_p} C_p \xleftarrow{\partial_{p+1}} C_{p+1} \xleftarrow{\partial_{p+2}}.$$

The relation $\partial_p \circ \partial_{p+1} = 0$ implies that the image of the boundary map ∂_{p+1} in C_p lies in the kernel of the boundary map ∂_p: $C_p \to C_{p-1}$. The quotient of these two spaces is used to define the pth-homology module H_p;

$$H_p = \frac{Ker\partial_p}{Im\partial_{p+1}}.$$

If there is a homeomorphism $\varphi: X \to Y$, with $\sigma = \varphi^{-1}: Y \to X$, the boundary function;

$$\hat{\partial}_p = \partial_p \circ \sigma = \sum_{j=0}^{j=p} (-1)^j f_{ij}^{p-1} \circ \sigma$$

acts on the chain complex;

$$0 \xleftarrow{\hat{\partial}_0} \hat{C}_0 \xleftarrow{\hat{\partial}_1} \hat{C}_1 \ldots \hat{C}_{p-1} \xleftarrow{\hat{\partial}_p} \hat{C}_p \xleftarrow{\hat{\partial}_{p+1}} \hat{C}_{p+1} \xleftarrow{\hat{\partial}_{p+2}}$$

where a typical element of $\hat{C}_p$ is a sum of elements of the form $n_i e_i^p \circ \sigma$ with n_i a positive integer.

We now follow the Alexander–Kolmogorov (A–K) construct for a cohomology dual to the homology we have now built.

Recall that the topological invariant C_p is the freely generated abelian group on the set of all p-embeddings e_i^p. A typical element of C_p is a sum of elements of the form $n_i e_i^p$. Define $HOM(C_p, T)$ to be the set of all homomorphisms $\varphi : C_p \to T$ where T is the translational (abelian) subgroup of the Poincaré group. Following the A–K construct, $HOM(_*, T)$ is a contravariant functor transforming the homology chain complex;

$$0 \xleftarrow{\partial_0} C_0 \xleftarrow{\partial_1} C_1 \ldots C_{p-1} \xleftarrow{\partial_p} C_p \xleftarrow{\partial_{p+1}} C_{p+1} \xleftarrow{\partial_{p+2}}.$$

Into the cohomology cochain complex;

$$0 \xrightarrow{\partial^0} HOM(C_0, T) \xrightarrow{\partial^1} HOM(C_1, T) \ldots HOM(C_{p-1}, T)$$
$$\xrightarrow{\partial^p} HOM(C_p, T) \xrightarrow{\partial^{p+1}} \cdots$$

For if $\varphi \in HOM(C_{p-1}, T)$ then $\partial^p(\varphi) \equiv \varphi \circ \partial_p \in HOM(C_p, T)$.

Identifying *embdX* as tessellated space–time let S be a step across the computational spin network from one node to another. S is the vector sum of a set of minimal steps from a node to a neighboring node of the random network. Each such minimal step is of the following form. It is a step from node $n(k)$ to a neighboring node $n(l)$ along the boundary edge of a p-simplex Δ_i^p for some value of $p \geqslant 2$. Thus we can choose a local homomorphism φ in the cochain complex $HOM(C_{p-1}, T)$ with $\varphi(S) = T$, an element of the translational subgroup of the Poincaré group. 'Glueing' these local homomorphisms together constitutes the desired mapping.

This result is a fundamental characterization of space–time. At the Planck scale, space–time is granular.

2.5 Quantum paths in space–time

We will define fibre bundles and the action of Lie groups on fibre bundles as gauge groups of automorphisms in more depth at the beginning of chapter 3. For now we simply note that a fibre *bundle* consists of a triple (E, B, π) with E a set, B a subset of E called the *base space* and *π a surjection from E onto B*. For $b \in B$, *the fiber rooted at b* is the set $\{\pi^{-1}(b)\} \cap E$. A *section* through a bundle is a mapping φ which is a continuous right inverse for the mapping π. This means that $\pi \circ \varphi$ is the identity mapping. We next show there is a local mapping from the translational subgroup T of the Poincaré group into a neighbourhood of the identity of a quantum Weyl algebra fibre bundle, such that the whole classical path can be lifted into the fibre bundle to form a unique quantum field as a section through the fibres.

In chapter 1 we defined a black box model of a quantum system in terms of an *information mapping* between inputs and outputs. We can apply the same approach to defining a classical dynamical system as an information mapping which locally reduces the system level *entropy*, a concept we will now define. Consider a classical system's phase space. The emergent behavior of this classical system gives rise to regions of phase space each corresponding to similar macro-level behaviour. The number and variation in size of these regions reflects the overall complexity of the system. The *entropy* of such a coarse-grained region is a count of all the different micro-configurations constituting that region. A system in which neither energy nor information can be exchanged across its boundary is *isolated.* An isolated system solution starting in a low entropy state will tend to wander into larger coarse-grained regions of phase space; hence entropy tends to increase. This effect gives rise to the second law of thermodynamics. The structure of classical phase space is such that each set of initial conditions (x_μ, p_μ) generates a unique solution; solutions do not cross. For a Hamiltonian system it is thus possible to reformulate classical mechanics as a symplectic vector space of solutions, or an equivalent set of initial conditions of location and momentum, equipped with a bilinear form Ω which ultimately derives from Hamilton's equations of motion as follows.

With $\xi \equiv (x, p)$ a six-dimensional solution vector we have;

$$\frac{\partial \xi}{\partial t} = \Omega \frac{\partial H}{\partial \xi} \text{ where } \Omega = \begin{bmatrix} 0_{3\times 3} & I_{3\times 3} \\ -I_{3\times 3} & 0_{3\times 3} \end{bmatrix}.$$

This is of the form of a symplectic *vector bundle* $V \oplus V^*$ (see below) where V is a real finite vector space with dual V^*. In a way similar to Hodge's reformulation of Maxwell's equations, the skew-symmetric rank 2 tensor Ω then takes the general form;

$$\Omega(x \oplus \eta, x' \oplus \eta') = \eta' x - \eta x'.$$

2.5.1 Fibre bundle structure of classical phase space

We take V to be the configuration space, V^* the (dual) momentum space and $V \oplus V^*$ the phase space. With this structure, phase space is a finite *product vector bundle* over V with fibre V^*. By choosing values such as (1,0,0,0,0,0) we can pull out specific elements;

$$\Omega(x \oplus \eta, 0 \oplus \eta') = \eta' x = \eta_1'.$$

From this point of view, the Dirac canonical quantisation of elements of phase space such as η_1' is equivalent to the canonical quantisation; $\Omega \to \hat{\Omega}$ as a (not necessarily bounded) linear operator, and this form of canonical quantisation extends smoothly to a countably infinite phase space [21].

Given this canonical quantisation; $\Omega \to \hat{\Omega}$, we can form the Weyl unitaries $\hat{W} = \exp i\hat{\Omega}$. Closure of linear combinations of these unitaries and their adjoints in the normed operator topology forms the *Weyl algebra* [21]. The extension of Ω to the space of solutions $\boldsymbol{S}$ allows us to define an inner product on $\boldsymbol{S}$ as $\langle S(1), S(2)\rangle = -i\Omega(S(1)^*, S(2))$ where $S(1)^*$ is the complex-conjugate solution. This defines an inner product on $\boldsymbol{S}$ relative to which a Hilbert space can be defined. For a quantum system of bosonic harmonic oscillators, we can then assemble a symmetric tensor product Fock space using Dirac's construction [22] based on creation and annihilation operators.

2.5.2 An example of the Weyl form

An example of the existence of the Weyl form in a two-dimensional locally flat space–time is now given for a local algebra $O(D)$, having a representation as observables acting on the Hilbert space $L^2(x, t)$ with Lebesgue measure.

For a small increment of space–time $(\delta x, \delta t)$ we consider the Poincaré translation subgroup element $T(\delta x, \delta t)$: $(x, t) \to (x + \delta x, t + \delta t)$ and define the mapping;

$$U_{T(\delta x,\, \delta t)}f(x, t) = f(x - \delta x, t - \delta t) \text{ for } f(x, t) \in L^2(x, t).$$

Then U is a local group homomorphism of the translation group T as observables acting on $L^2(x, t)$. Define also;

$$V_{\delta t} = U_{T(0,\, \delta t)};\ V_{\delta x} = U_{T(\delta x,\, 0)}$$
$$\Rightarrow V_{\delta t}V_{\delta t'}f(x, t) = U_{T(0,\, \delta t)}U_{T(0,\, \delta t')}f(x, t) = f(x, t - \delta t' - \delta t)$$
$$= U_{T(0,\, \delta t'+\delta t)}f(x, t) = V_{\delta t+\delta t'}f(x, t) \Rightarrow V_{\delta t}V_{\delta t'} = V_{\delta t+\delta t'}.$$

A similar result applies for $V_{\delta x}$, by symmetry.

Now introduce a deformation of the form; $T(\delta x, 0) \to Z_{\delta x}f(x, t) = \exp(it\delta x)f(x - \delta x, t)$. The mappings V and Z are unitary representations on $L^2(x, t)$ and so also is their product

$$V \times Z : T(\delta x, \delta t) \to V_{\delta t}Z_{\delta x}.$$

We have;

$$\begin{aligned}
V_{\delta t}Z_{\delta x}f(x, t) &= V_{\delta t}\exp(it\delta x)f(x - \delta x, t)\\
&= \exp(it\delta x)f(x - \delta x, t - \delta t)\\
Z_{\delta x}V_{\delta t}f(x, t) &= Z_{\delta x}f(x, t - \delta t)\\
&= \exp(i(t - \delta t)\delta x)f(x - \delta x, t - \delta t)\\
&= \exp(-i\delta t\delta x)V_{\delta t}Z_{\delta x}f(x, t)\\
Z_{\delta x}V_{\delta t} &= \exp(-i\delta t\delta x)V_{\delta t}Z_{\delta x}.
\end{aligned}$$

Then $T(\delta x, \delta t) \to (Z_{\delta x}V_{\delta t})$ is a local Weyl representation of the CCR on $L^2(x, t)$. By the Stone–von Neumann theorem, the resulting C*-algebra and its weak closure as a von Neumann algebra must be unitarily isomorphic to Wald's equivalent 'algebraic approach' to quantum field construction and his Weyl algebra, since we can assume all relevant Hilbert spaces are either finite or separable in application to observed real systems. Any two separable Hilbert spaces are unitarily equivalent, as are two quantum systems of the same finite energy and dimensionality.

This example indicates that a continuous local group homomorphism from a neighbourhood of the identity of T to a neighbourhood of the identity of the set of observables in $O(D)$ exists as a Weyl algebra. It can be easily extended to four dimensions by replacing x by the 3-vector $\mathbf{x} = (x_1, x_2, x_3)$.

A discrete path in space–time, as we have so far generated it, can be considered as a set of linked causally directed vectors in space–time. More formally, we define the path as a series of n linked increments $\mathbf{a}(j)$ each with varying direction within their local forward light cones, such that the path begins at $\mathbf{x(0)}$ and ends at $\mathbf{x(1)}$, with $T(a(j)){:}x \to x + a(j)$ elements of the translation subgroup $\boldsymbol{T}$. The total path is then generated by the product $\prod_{j=1}^{j=n} T(a(j))\boldsymbol{x}(\mathbf{0})$ with the final end point $\mathbf{x}(1) = \boldsymbol{x}(\mathbf{0}) + \sum_{j=1}^{j=n} a(j)$. For a fixed initial point $\boldsymbol{x}(\mathbf{0})$ we can identify this path with the finite group product $\prod_{j=1}^{j=n} T(a(j)) \in \boldsymbol{T}$. If all the increments point in the same direction, and the local light cones have the same orientation, one to another, this gives a straight path in space–time. If, however, the local light cones are rotated one to another, then the path will be curved and the curvature will be due to the existence of a gravitational field. The choice of settings for the light cones constrains the trajectory of the resultant path in the same way that the gravitational field in Einstein's theory constrains a geodesic path between two spacetime events to be of the form

$$\frac{d^2x_\tau}{ds^2} + \Gamma^\tau_{\mu\nu}\frac{dx_\mu}{ds}\frac{dx_\nu}{ds} = 0$$

where, in Christoffel notation, $\Gamma^\tau_{\mu\nu} = \{\mu\nu, \tau\}$, the *Christoffel tensor*. For the quantum case a reorientation of the locally compact light cones resulting in a different gravitational field corresponds to an isomorphism between a path and a finite set of such orientations, one for each node of our network, assuming Lorentzian causality.

The orientation of these locally compact light cones in classical space–time corresponds to the definition of three angles (the pitch, roll and yaw angles) as a function of time t. We can capture these parameters by the general quaternion $=t + i\theta + j\varphi + k\sigma$. In the quantum case we can use Dirac 4-spinors, giving rise to discrete solutions.

Now let there be given a continuous local group homomorphism from a neighbourhood V of the identity of T to the neighbourhood W of the identity of the set of observables in $O(D)$ as a Weyl representation of the CCR. Then a discrete classical path CP waymarked in space–time as described by oriented locally compact light cones can be lifted to a quantum field section QP through the bundle $O(D)$.

Proof. We construct the section iteratively, following a method suggested by Pontryagin [23]. Let CP be a discrete classical path. From the definition, the path is a series of n linked vectors $\mathbf{a}(j)$ each of Euclidean interval length $|\mathbf{a}(j)|$ and varying direction such that the path begins at $\mathbf{x(0)}$ and ends at $\mathbf{x(1)}$, with $T(\mathbf{a}(j))$: $\mathbf{x} \to \mathbf{x} + \mathbf{a}(j)$ elements of the translational subgroup $\boldsymbol{T}$ of the Poincaré group.

We have, by assumption, a continuous local group homomorphism φ: $T(a) \to U_{T(a)}$ from V, a neighbourhood of the identity of $\boldsymbol{T}$ to W, a neighbourhood of the identity of $(O(D))$. Clearly, by redefining the number of links in CP if necessary, we can assume that $|\mathbf{a}(j)|$ is sufficiently small so that $T(\mathbf{a}(j)) \in V$ for all j and by choosing appropriate units we can assume that CP consists of n links each of length $1/n$. This is similar to the characterization earlier in the chapter concerning steps across space–time as composed of a set of minimal steps. Thus, this argument works in the regime of granulated space–time.

We additionally assume that for n large;

$$|\mathbf{x}(t_2) - \mathbf{x}(t_1)| \ < \tfrac{1}{n} \Rightarrow T(\mathbf{x}(t_1))^{-1}T(\mathbf{x}(t_2)) = T(\mathbf{a}(t_1)) \in V.$$

Let m be a positive integer strictly less than n and suppose that the quantum path QP has been defined such that its initial value is $QP(0) = \mathbf{A(x(0))}$. We proceed by induction. Assume that QP has been defined for all values of $\mathbf{x}(t)$ with $0 \leqslant t \leqslant \frac{m}{n}$, and satisfies, for all such t with $0 \leqslant t \leqslant \frac{m}{n}$;

(a) the fixed endpoint assumption; $QP(0) = \mathbf{A(x(0))}$;

(b) the local lifting assumption to the Weyl algebra near the identity; If $\mathbf{x}(s)$, $\mathbf{x}(t)$ in CP satisfy $|\mathbf{x}(s) - \mathbf{x}(t)| \ \leqslant \frac{1}{n}$ then $T(x(s))^{-1}T(x(t)) \in V$ and $\varphi(T(x(s))^{-1}T(x(t)) = QP(x(s))^{-1}QP(x(t)) \in W$.

We now extend the path $\{QP(\mathbf{x}(t)); 0 \leqslant t \leqslant \frac{m}{n}\}$, stepping forward one additional link on CP so that $t = \frac{m}{n} + \frac{1}{n}$; using the following construction;

$$QP\left(\mathbf{x}\left(\frac{m+1}{n}\right)\right) = QP\left(\mathbf{x}\left(\frac{m}{n}\right)\right)\varphi\left(T\mathbf{x}\left(\frac{m}{n}\right)^{-1}T\mathbf{x}\left(\frac{m+1}{n}\right)\right). \tag{2.1}$$

From equation (2.1) the extension of the path QP still satisfies

(a): $QP(0) = \mathbf{A(x(0))}$ since φ acting on the identity of the group of local translations $\boldsymbol{T}$ is the identity operator in $O(D)$. We need to show that the extension under induction still satisfies the local lifting assumption (b).

Let h be a real number with $|h| \ \leqslant \frac{1}{n}$. If h is positive, then by definition h satisfies the extension shown in equation (2.1). Thus, we have;

$$QP\left(\mathbf{x}\left(\frac{m}{n} + h\right)\right) = QP\left(\mathbf{x}\left(\frac{m}{n}\right)\right)\varphi\left(T\mathbf{x}\left(\frac{m}{n}\right)^{-1}T\mathbf{x}\left(\frac{m}{n} + h\right)\right).$$

If on the other hand h is negative then setting $x(s) = x(\frac{m}{n} + h)$ and $x(t) = x(\frac{m}{n})$; since s, t are both equal to or less than $\frac{m}{n}$; by the inductive hypothesis they therefore satisfy;

$$\varphi((Tx(s))^{-1}Tx(t)) = (QPx(s))^{-1}QPx(t) \text{ with } t = \frac{m}{n} + h \text{ and } s = \frac{m}{n}$$

$$\Rightarrow \varphi\left(\left(Tx\left(\frac{m}{n}\right)\right)^{-1} Tx\left(\frac{m}{n} + h\right)\right) = \left(QPx\left(\frac{m}{n}\right)\right)^{-1} QPx\left(\frac{m}{n} + h\right)$$

$$\Rightarrow QPx\left(\frac{m}{n} + h\right) = QPx\left(\frac{m}{n}\right)\varphi\left(Tx\left(\frac{m}{n}\right)^{-1} Tx\left(\frac{m}{n} + h\right)\right).$$

Thus equation (2.1) holds for both positive and negative values of h.
It follows that;

$$\left(QP\mathbf{x}\left(\frac{m+1}{n}\right)\right)^{-1} QP\mathbf{x}\left(\frac{m}{n} + h\right) = \left(QP\mathbf{x}\left(\frac{m}{n}\right)\varphi\left(T\mathbf{x}\left(\frac{m}{n}\right)^{-1} T\mathbf{x}\left(\frac{m+1}{n}\right)\right)\right)^{-1}$$
$$\times\left(QP\mathbf{x}\left(\frac{m}{n}\right)\varphi\left(T\mathbf{x}\left(\frac{m}{n}\right)^{-1} T\mathbf{x}\left(\frac{m}{n} + h\right)\right)\right)$$
$$= \varphi\left(T\mathbf{x}\left(\frac{m+1}{n}\right)^{-1} T\mathbf{x}\left(\frac{m}{n} + h\right)\right).$$

Thus, the path extension satisfies both requirements (a) and (b) completing the inductive step $\frac{m}{n} \to \frac{m+1}{n}$, provided we satisfy the local topological constraints, namely;

We have that if n is sufficiently large then there is a neighbourhood U such that; $|x(t_2) - x(t_1)| < \frac{1}{n} \Rightarrow$ by continuity $T(x(t_1))^{-1}T(x(t_2) = T(a(t_1)) \in U^{-1}U \subset V$ and $\varphi(T(x(t_1))^{-1}T(x(t_2)) = \varphi(T(a(t_1)) \in W$

Setting, for small $h>0$;

$$t_1 = \frac{m+1}{n};\ t_2 = \frac{m}{n} + h \Rightarrow |x(t_1) - x(t_2)| \text{ small}$$
$$\Rightarrow Tx\left(\frac{m+1}{n}\right)^{-1} Tx\left(\frac{m}{n} + h\right)$$
$$= T(x(t_1))^{-1}T(x(t_2)) = T(x(t_2))T(x(t_1))^{-1}$$
$$= T(x(t_2) - x(t_1)) \in V$$
$$\varphi\left(T\left(x\left(\frac{m+1}{n}\right)\right)^{-1} Tx\left(\frac{m}{n} + h\right)\right) \in W \text{ as required.}$$

Since φ (identity of T) = identity of $O(D)$ the induction hypothesis is true for $\frac{m}{n} = 0$; $h = \frac{1}{n}$. By induction the path $QP(t)$ can be extended in $O(D)$ for all discrete steps m less than or equal to n.

We now show that the constructed quantum field QP is unique.

Proof. The initial point of QP is unique by condition (a).

If $QP(\mathbf{x}(\mathrm{t}))$ is unique for all $t \leqslant t_0$ then let $t_0 < t \leqslant t_0 + \varepsilon$. Then;

$$T\mathbf{x}(t_0)^{-1}T\mathbf{x}(t) \in V \Rightarrow \varphi(T\mathbf{x}(t_0)^{-1}T\mathbf{x}(t)) = QP\mathbf{x}(t_0)^{-1}QP\mathbf{x}(t) \Rightarrow QP\mathbf{x}(t) \\ = QP\mathbf{x}(t_0)\varphi(T\mathbf{x}(t_0)^{-1}T\mathbf{x}(t)).$$

Thus, the path QP is uniquely determined for all points $t < t_0 + \varepsilon$. The result follows by induction.

Finally, we show that there is a projection π from the fiber bundle $O(D)$ mapping the quantum field back to the path CP.

Proof. From condition (a) this is clear for the initial point. We again use induction to prove the general case. If the projection π maps the observable $QP\ \mathbf{x}(t)$ back to $T\mathbf{x}(t)$ for $t \leqslant t_0$, let $t_0 < t \leqslant t_0 + \varepsilon$. Then as before we have;

$$QP\mathbf{x}(t) = QP\mathbf{x}(t_0)\varphi(T(\mathbf{x}(t_0))^{-1}T\mathbf{x}(t))$$
$$\text{thus } \pi QP\mathbf{x}(t) = \pi QP\mathbf{x}(t_0)\pi\circ\varphi(T(\mathbf{x}(t_0))^{-1}T\mathbf{x}(t)).$$
$$= T\mathbf{x}(t_0)(T(\mathbf{x}(t_0))^{-1}T\mathbf{x}(t)) = T\mathbf{x}(t)$$

The result follows by induction.

2.6 Fractal paths in classical space–time

The principle of relativity is captured within the assumptions of the Riemannian geometry of 4-manifolds, where formulae equating a tensor expression to zero remain covariant under local diffeomorphism transformations. It is a natural extension of these ideas to additionally postulate that the scales of measurement inscribed on the clocks or measuring rods used by an observer should also not be absolute. Mathematically this can be captured by the additional requirement that the tensor formulae should be invariant under transformations of scale. From this perspective a classical relativistic system is a scale free' system. However, if space-time is tessellated then this provides a lower limit to scale transformations and similar renormalisation processes such as that of Gell-Mann and Low [24]. *Thus, our theory of algebraic quantum gravity is both finite, in the sense of Dirac, and non-fractal.*

By contrast, assume that we have a fractal closed loop L in classical smooth space–time with Euclidean diameter d following Wick rotation. We approximate L by a discrete closed path $L(\eta)$ where η is the Euclidean length of each segment of $L(\eta)$. We will prove that such a loop is renormalisable with limit equal to the loop fractal dimension.

Proof. Standard dimensional analysis shows that $N(\eta)$, the number of segments in the path $L(\eta)$, is a function of the form $f(\frac{d}{\eta})$. We will establish the nature of this function and its renormalisation limit. The fractal, self-similar nature of the discrete path implies that if we consider a finer segmentation of segment length ξ, then;

$$N(\xi) \propto N(\eta)N(\xi|\eta) = \frac{f\left(\frac{d}{\eta}\right)f\left(\frac{\eta}{\xi}\right)}{f(1)^2}$$

where $N(\xi|\eta)$ is the number of segments of length ξ in a segment of length η, and $N(d) = N(\eta|\eta) = f(1)$.

It follows that $\frac{f(\frac{d}{\xi})}{f(1)} = \frac{f(\frac{d}{\eta})f(\frac{\eta}{\xi})}{f(1)^2}$ thus $f(\frac{d}{\xi}) = \frac{f(\frac{d}{\eta})f(\frac{\eta}{\xi})}{f(1)}$.

This implies that f for the limiting case, must be of the form $f\left(\frac{x}{y}\right) = C\left(\frac{x}{y}\right)^D$ with C and D constants; $C = f(1)$. Thus, we have;

$$f\left(\frac{d}{\eta}\right) = f(1)\left(\frac{d}{\eta}\right)^D \Rightarrow N(\eta) = \left(\frac{d}{\eta}\right)^D \text{and } L(\eta) = \eta\left(\frac{d}{\eta}\right)^D$$

Locally this implies $L(\xi|\eta) = \xi(\frac{\eta}{\xi})^D = \eta^D\xi^{1-D}$.

The renormalisation limit is thus finite and we identify D with the path fractal dimension.

In the final part of this chapter, in preparation for chapter 3, we discuss how supersymmetry and Dirac's relativistic spinor calculus interact.

2.7 Supersymmetry and the spinor calculus

We can generalise spinors as $n \times n$ matrices concentrating on the case n even so that $n = 2l$, $(l = 1,2)$.

In this context we define the generalised *Dirac operator* $\hat{D}$, assuming classical flat space–time, by;

$$\hat{D} = \sum_{j=0}^{j=2l-1} \gamma_j \frac{\partial}{\partial x_j} \Leftrightarrow \hat{D}^2 = -\Delta \Leftrightarrow (i\hat{D})^2 = \Delta = \sum_{j=0}^{j=2l-1} \frac{\partial^2}{\partial x_j^2}.$$

There are intriguing connections between Atiyah's index theorem applied to the Dirac operator, the corresponding Todd polynomial and work by Donaldson, Sieberg and Witten relevant to the homology invariants of four-dimensional space–time [25].

We introduced the cases $l = 1, 2$ in chapter 1 informally, when discussing solutions to the Dirac equation. We focus here on these cases.

2.7.1 2-Spinors

Let $SL(2, C)$ be the special linear group of 2×2 matrices with complex entries whose determinant equals 1. Then $SL(2, C)$ has two inequivalent spinor representations:

(a) The self-representation $M \to M$ and

(b) The complex-conjugate self-representation $M \to M^*$.

The self-representation consists of matrices M acting on a two-dimensional vector space F whose canonical basis $\{e_1, e_2\}$ is required to consist of two *Clifford variables.* This means that

$$e_i e_j = -e_j e_i \text{ for } i, j = 1, 2, i \neq j.$$

Similarly, the complex-conjugate self-representation consists of matrices M^* acting on a two-dimensional space denoted $\dot{F}$, whose canonical basis $\{f_1, f_2\}$ also consists of Clifford variables.

A *2-spinor* ψ is any element of the vector space F. It is denoted ψ_A $(A = 1,2)$ and called a left-handed Weyl 2-spinor. Given such a spinor ψ_A, the matrix M acting on ψ_A gives rise to another spinor $\psi'_A = M_A^B \psi_B$ assuming the Einstein summation convention. This index notation introduced by H Weyl forms the basis of the *spinor calculus* notation. In this spinor notation, elements of the space $\dot{F}$ are denoted $\bar{\psi}_{\dot{A}}$ and called right-handed Weyl 2-spinors. Left-handed and right-handed Weyl spinors are related as follows;

$$\psi_A \in F \text{ implies } \psi_A^* = \bar{\psi}^A \in F^*.$$

The representations $M \to M$ and $M \to (M^{-1})^T$ are unitarily equivalent, so there is a unitary matrix $\varepsilon = \varepsilon_{AB}$ such that for all M, $\varepsilon M \varepsilon^{-1} = (M^{-1})^T$ where $\varepsilon = \begin{pmatrix} 0 & -1 \\ 1 & 0 \end{pmatrix}$.

Defining $\varepsilon^{AB} = \varepsilon_{AB}^{-1}$ we have $\varepsilon^{AB} = \begin{pmatrix} 0 & 1 \\ -1 & 0 \end{pmatrix}$. The matrices ε_{AB} and ε^{AB} are tensor objects which can be used to raise and lower indices in the usual vector and tensor calculus, within which left-handed Weyl 2-spinors are covariant vectors, and right-handed Weyl 2-spinors are contravariant vectors.

We similarly define the matrix $\bar{\varepsilon}$ such that the representations $M \to M^*$ and $M \to (M^{*-1})^T$ are equivalent, so that $\bar{\varepsilon} M^* \bar{\varepsilon}^{-1} = (M^{*-1})^T$ for all M.

Then $\bar{\varepsilon} = \varepsilon^{\dot{A}\dot{B}} = \begin{pmatrix} 0 & 1 \\ -1 & 0 \end{pmatrix}$ and $\bar{\varepsilon}^{-1} = \varepsilon_{\dot{A}\dot{B}} = \begin{pmatrix} 0 & -1 \\ 1 & 0 \end{pmatrix}$.

We then have the following vector/tensor relationships for example;

$$\psi^A = \varepsilon^{AB}\psi_B = -\psi_B \varepsilon^{BA}; \quad \psi_A = \varepsilon_{AB}\psi^B$$
$$\bar{\psi}^{\dot{A}} = \varepsilon^{\dot{A}\dot{B}}\bar{\psi}_{\dot{B}} = -\bar{\psi}_{\dot{B}}\varepsilon^{\dot{B}\dot{A}}; \; \bar{\psi}_{\dot{A}} = \varepsilon_{\dot{A}\dot{B}}\bar{\psi}^{\dot{B}}.$$

Since the spinor representation space F is a vector space, we can consider the dual space F^* of F. We then have the following commutative diagram (figure 2.1);

Where M maps $F \to F$ by $\psi'_A = M_A^B \psi_B$ and $(M^{-1})^T$ maps $F^* \to F^*$ by $\psi'^A = \varepsilon^{AB}\psi'_B = \varepsilon^{AB} M_B^C \varepsilon_{CD} \psi^D = (M^{-1})^T{}^A_D \psi^D$.

For basis elements $\psi \in F^*$ and $\chi \in F$ we can consider entangled (tensor) products such as $\psi\chi = \psi^A \chi_A = \varepsilon^{AB}\psi_B \chi_A = \varepsilon^{12}\psi_2\chi_1 + \varepsilon^{21}\psi_1\chi_2 = \psi_2\chi_1 - \psi_1\chi_2$.

The Clifford algebra assumptions thus imply that products such as $\psi\psi$ are not trivially zero.

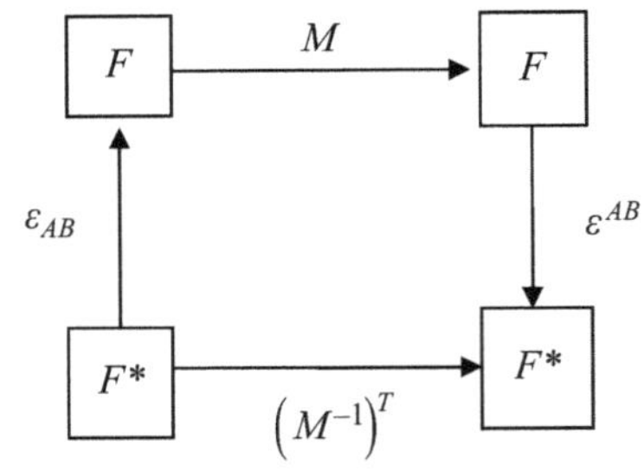

Figure 2.1. The dual space and corresponding mapping.

2.7.2 4-Spinors and the Lorentz and Poincaré groups

In chapter 1 we introduced the idea of special relativity based on the fact that the speed of light c is invariant with respect to the speed of the source. Thus in an inertial reference frame (that is, a reference frame moving at constant velocity) we have, applying Pythagoras' theorem, in time interval Δt, $c\Delta t = \sqrt{\Delta x^2 + \Delta y^2 + \Delta z^2}$ or $c^2\Delta t^2 = \Delta x^2 + \Delta y^2 + \Delta z^2$. Write the 4-vector $x^\mu = \{ct, x, y, z\}$ as an element of Minkowski space–time, and the *metric tensor*

$$\eta_{\mu\nu} = 0(\mu \neq \nu),\ \eta_{\mu\nu} = -1(\mu = \nu = 0),\ \eta_{\mu\nu} = 1(\mu = \nu = 1, 2, 3).$$

We then have the following; $\eta_{\mu\nu}x^\mu x^\nu = \eta_{\mu\mu}(x^\mu)^2 = 0$.

Transforming to a different inertial reference frame x'^μ leaves the speed of light invariant, thus $\eta_{\mu\mu}x'^\mu x'^\mu = 0$. The *Lorentz group* $\boldsymbol{L}$ is the group of linear transformations $\{\Lambda^\mu_\nu : x^\nu \to x'^\mu\}$ which respect this invariance (i.e. which leave the speed of light invariant in both reference frames). Thus;

$$(x')^2 = x'^\mu x'^\mu = \eta_{\mu\nu}x'^\mu x'^\nu = \eta_{\mu\nu}\Lambda^\mu_\rho x^\rho \Lambda^\nu_\tau x^\tau.$$

However, we also have, from invariance of the speed of light, $(x')^2 = (x)^2 = \eta_{\rho\tau}x^\rho x^\tau$. Equating these two we then have $\eta_{\mu\nu}\Lambda^\mu_\rho\Lambda^\nu_\tau = \eta_{\rho\tau}(\mu, \nu, \rho, \tau = 0, 1, 2, 3)$.

The metric tensor $\eta_{\mu\nu}$ is $\begin{pmatrix} -1 & 0 & 0 & 0 \\ 0 & +1 & 0 & 0 \\ 0 & 0 & +1 & 0 \\ 0 & 0 & 0 & +1 \end{pmatrix}$.

Across a finite volume of *curved* space–time, the $\eta_{\mu\nu}$ become functions, with respect to an arbitrarily chosen set of coordinates, that describe the metrical relations of space–time corresponding to the existence of a gravitational field.

As with the spinor calculus, where we used the expression ε_{AB} to raise and lower spinor indices, we can use the metric tensor $\eta_{\mu\nu}$ to raise and lower space–time indices. Following the spinor calculus we define $\eta^{\mu\nu}$ to be the matrix inverse of $\eta_{\mu\nu}$; which for the metric tensor in flat space–time clearly equals $\eta_{\mu\nu}$.

We define the restricted Lorentz group in such a way that $\det(\Lambda^\mu_\nu) = 1$ and $\Lambda^0_0 \geqslant 1$. This group is denoted $\boldsymbol{L}^\uparrow_+$ (+ for determinant +1 and ↑ for $\Lambda^0_0 \geqslant 1$). In general, we can have $\det(\Lambda^\mu_\nu) = \pm 1$, and $(\Lambda^0_0)^2 \geqslant 1$, which defines the full Lorentz group $\boldsymbol{L}$ of all orthogonal transformations of this type, denoted $O(1,3)$. The restricted Lorentz group $\boldsymbol{L}_+^\uparrow$,which satisfies the physics constraints, is a subgroup of $\boldsymbol{L}$. The full Lorentz group $\boldsymbol{L}$ is a non-compact Lie group. Thus any element of $\boldsymbol{L}$ (and hence also any element of $\boldsymbol{L}_+^\uparrow$) is of the form $\Lambda = \exp(tM)$ where t is a parameter and M, (the *generator* of Λ) is a linear transformation, and is an element of the corresponding Lie algebra. This Lie algebra in fact consists of 4×4 matrices with real entries. As illustrated in chapter 1, the elements Λ^μ_ν of the Lorentz group are rotations of the four vector x^μ in flat space–time, and their corresponding Lie algebra generators $M_{\mu\nu}$ become the quantum operators corresponding to angular momentum. Thus we can see that the properties of invariance in flat space–time are intimately connected to the properties of quantum operators.

If we consider such a rotation in two dimensions, then we have, for a rotation θ;

$$\begin{pmatrix} x_1 \\ x_2 \end{pmatrix}' = \begin{pmatrix} \cos\theta & \sin\theta \\ -\sin\theta & \cos\theta \end{pmatrix}\begin{pmatrix} x_1 \\ x_2 \end{pmatrix}.$$

Assuming θ is small we then have (ignoring elements of order θ^2 and higher);

$$\begin{pmatrix} x_1 \\ x_2 \end{pmatrix}' = \begin{pmatrix} 1 & \theta \\ -\theta & 1 \end{pmatrix}\begin{pmatrix} x_1 \\ x_2 \end{pmatrix} = \begin{pmatrix} 1 & 0 \\ 0 & 1 \end{pmatrix}\begin{pmatrix} x_1 \\ x_2 \end{pmatrix} + \begin{pmatrix} 0 & \theta \\ -\theta & 0 \end{pmatrix}\begin{pmatrix} x_1 \\ x_2 \end{pmatrix}.$$

Writing $\delta x^\mu = x'^\mu - x^\mu$ we can express this as;

$\delta x^\mu = \omega^\mu_\nu x^\nu$ where ω^μ_ν is the matrix $\begin{pmatrix} 0 & \theta \\ -\theta & 0 \end{pmatrix}$.

Note that the matrix ω^μ_ν is anti-symmetric, i.e. $\omega^\mu_\nu = -\omega^\nu_\mu$.

Also, $\omega^\mu_\nu = \theta\begin{pmatrix} 0 & 1 \\ -1 & 0 \end{pmatrix} = -i\theta\begin{pmatrix} 0 & i \\ -i & 0 \end{pmatrix} = -i\theta\sigma^1$ where σ^1 is one of the Pauli spin matrices. This again illustrates the intimate linkage between rotations in space-time and quantum operators.

For a large rotation θ we can write $R(\theta) = (R(\frac{\theta}{n}))^n$ so that rotation though θ is the product of a number of very small rotations.

For $\frac{\theta}{n}$ a small angle, we have, from above, $x'^\mu = x^\mu + \omega^\mu_\nu x^\nu = x^\mu - \frac{i\theta}{n}\sigma_1 x^\mu = x^\mu\left(I - \frac{i\theta}{n}\sigma_1\right)$

where I denotes the 2×2 identity matrix.

For a rotation of θ we have $x'^\mu = x^\mu(I - \frac{i\theta}{n}\sigma_1)^n \to x^\mu \exp(-i\theta\sigma_1)$ as $n \to \infty$.

Thus the Lie group action of rotation by θ in two-dimensional space–time is generated by $\exp(-i\theta\sigma^1)$ where σ^1 is the generating spin operator in the corresponding Lie algebra.

Similarly, a Lorentz transformation, considered as a rotation in four-dimensional space–time is of the form $\Lambda = I + \omega$ with ω a linear transformation. These operators ω correspond to either rotation around each of the three space axes, or Lorentz boosts resulting in a change in velocity along each of the three space axes. They are denoted respectively by $J_i(i = 1, 2, 3)$ for the rotations, and $K_i(i = 1, 2, 3)$ for the boosts. These two sets of linear transformations are generically denoted by the abstract tensor $M_{\mu\nu}$. In this sense, the elements ω of the Lie algebra, with ω 'small', can be considered as members $(I + \omega)$ of the tangent space of the identity I of the Lorentz group.

In addition to rotation about each of the three space axes x, y and z, we can also consider a general translation in space time of the form $x'^{\mu} = x^{\mu} + a^{\mu}(\mu = 0, 1, 2, 3)$. Such a translation with a^{μ} constant clearly preserves the speed of light. Adding this operation to the Lorentz group of transformations gives the *Poincaré group*. This then consists of all transformations of space–time of the form $x'^{\mu} = \Lambda^{\mu}_{\nu}x^{\nu} + a^{\mu}$. The Poincaré group is also a Lie group, with the Lorentz group as a proper subgroup. The Lie algebra generators of these translations a^{μ} in space–time are denoted by the elements of the abstract vector P_{μ} $(\mu = 0, 1, 2, 3)$.

2.7.3 4-Spinors and the spin groups

The case $n = 4$ $(l = 2)$ corresponds to the Dirac 4-spinors γ^{μ} which we came across in chapter 1 as part of the solution to the Dirac equation. To bring them even closer to a representation of flat space–time we can tweak them so that they satisfy $\{\gamma^{\mu}, \gamma^{\nu}\} = \eta^{\mu\nu} = \eta_{\mu\nu}=$ the metric tensor or its inverse. For a^{μ} ranging over all real 4-vectors, define the real vector space;

$$V = \{\gamma.\ a\} = \{\eta_{\mu\mu}\gamma^{\mu}a^{\mu}\}$$

Let S be the group of all non-singular matrices S implementing an automorphism on V as $\alpha_s(\gamma.\ a) = \{S(\gamma.\ a)S^{-1}\}$. *We define* S *to be the group Pin(1,3).*

It is easy to show that the Dirac spinors γ^{μ} are a basis for the vector space V. Thus, for some real 4-vector $a' = a'^{\mu}$ we have $S(\gamma.\ a)S^{-1} = \gamma.\ a'$. The mapping $a \to a'$ is linear and, in fact, this mapping implements an equivalent Lorentz transformation of real 4-vectors denoted $\wedge(S)$. In sum we have the following expression [26];

$$S(\gamma.\ a)S^{-1} = \gamma.\ a' = \gamma\cdot \wedge (S)a$$

The mapping $\mp S \to \wedge (S)$ identifies *Pin*(*1*,3) with the full Lorentz group $O(1,3)$. The same mapping identifies the subgroup *Spin*(1,3) with the Lorentz subgroup $L_{+} = SO(1,3)$. *Thus, we have a very close relationship between abstract spinor calculus and the symmetries of space–time.* To demonstrate further the close linkage between properties of space–time and the algebraic representation of spinors, we note that there is an isomorphism between the spinor representation of $SL(2,C)$ and the Lorentz group $L^{\uparrow}_{+}$. Thus the Lorentz group transformation Λ of space–time is echoed by a corresponding transformation ad$M$ which maps the 2×2 matrix $\pm X$ of $SL(2,C)$ to the matrix $MXM^{\dagger}$ of $SL(2, C)$. This works because any such matrix X can be written in the form; $X = x_{\mu}\sigma^{\mu}$ where x_{μ} is a point in space–time, and the σ^{μ} are the extended Pauli spin matrices.

2.8 Irreducible representations of the Poincaré Lie algebra

The Lie algebra generator P_μ has the *Casimir property* that its square $P^2 = P_\mu P^\mu$ lies in the *centre* of the Lie algebra. This means that as well as being in the algebra, it commutes with every other element of the algebra. It turns out that there is precisely one other operator which has the same property, namely the square W^2 of the Pauli–Ljubanski operator W. If we consider an irreducible representation of the Poincaré Lie algebra, this means that the algebra is represented as a set of operators acting on an underlying quantum state space, and implies that the centre of the algebra consists of multiples of the identity operator I. (Any non-trivial element of the centre of the algebra would allow the quantum state space to be split into two parts by a projection operator, contradicting the assumption of being an irreducible representation.)

In this irreducible representation we must then have both P^2 and W^2 as multiples of the identity since they are in the centre of the representation. Due to the various algebraic properties of P^2 and W^2 it results that they have the form;

$P^2 = m^2 I$ and $W^2 = -m^2 s(s+1)I$, where m is the mass and s is the spin of the corresponding particle. The black box, positivist, interpretation of this result is that an irreducible representation of the Poincaré–Lie algebra *is a unique particle* with mass m and spin s. We can consider a number of cases, as follows.

(a) The mass m is positive and the spin s is a fixed value which is either half integer (fermionic) or integer (bosonic) in character. This represents a single massive particle, either a fermion or a boson depending on the spin value.

(b) Both P^2 and W^2 are zero. In this case it turns out that $P_\mu = sW_\mu$. This irreducible representation corresponds to a massless particle of spin $\pm\, s$ such as a photon.

These two are the only cases which seem to arise in nature.

For the standard model (i.e. the special relativistic theory of elementary particles without supersymmetry), a particle is mathematically defined in terms of its mass m and spin s, by an irreducible representation of the Lie algebra of the Poincaré group. Such particles may also relate to other 'gauge type' symmetries. For example, photons are the carriers of the electromagnetic force and thus obey a $U(1)$ *gauge symmetry* consistent with that.

2.8.1 The supersymmetric extension of the standard model (SSM)

The idea of the SSM is to add additional vector elements, normally denoted $Q_{\alpha j}$, $j = 1, 2,\ldots$ to the set of abstract tensors and vectors $\{M_{\mu\nu}, P_\mu\}$ which form the Lie algebra of the Poincaré group. To ease notational clutter, we will often write these additional elements as simply Q_α. In any representation, these are all linear operators, including the identity operator I, and thus form a subset of an algebra of such operators. This is done in such a way that an irreducible such representation corresponds not to a single particle, but to a set of particles, called a *multiplet*, all of the same mass m, unless this symmetry is broken, and containing equal numbers of bosons (particles with integer spin values such as 0 or 1) and fermions (particles with half integer spin such as 1/2).

Mathematically the set $\{M_{\mu\nu}, P_\mu\} \cup \{Q_{\alpha j}\}$ no longer forms a standard Lie algebra, but in fact forms a more general object called a *graded Lie algebra*. Additional mathematical machinery is required to define multiplication in this enhanced algebra, resulting in the ability to mix together both commutators and anticommutators. By this means we subvert the restrictions of the Coleman–Mandula 'no free lunch' theorem which shows that a more straightforward Lie algebra-based extension of the Poincaré group with new physics is not possible; the various parts of the machine spin round but do not mesh. Part of this machinery is provided by the *Dirac spinors* and the corresponding *spinor calculus*. So, we need to define these first, before then constructing the corresponding graded Lie algebra.

2.9 Dirac spinors and the spinor calculus

Recall that a Poincaré transformation P is a Lorentz transformation Λ followed by a translation a in flat space–time. If x^μ ($\mu = 0, 1, 2, 3$) is a point in space–time, then;

$$x'^\mu = \Lambda^\mu_\nu x^\nu + a^\mu.$$

The generators of the Poincaré group as a Lie group are the six generators $M_{\mu\nu}$ ($\mu, \nu = 0, 1, 2, 3$) of the Lorentz group and the four generators P_μ ($\mu = 0, 1, 2, 3$) of translations in space–time. In representations of these as quantum operators, the $M_{\mu\nu}$ ($\mu, \nu = 0, 1, 2, 3$) correspond to angular momentum observables, and P_μ ($\mu = 0, 1, 2, 3$) correspond to energy observables. (Recall from chapter 1 that in quantum mechanics each observable corresponds to a quantum operator in a finite dimensional or separable representation; the eigenvalues of this operator are the observed values.)

We can develop a particular representation of these operators using the Dirac spinor matrices which arise from the Dirac equation. The Dirac equation describes the dynamics of the quantum state of a spin 1/2 fermion (such as an electron) in a way which is consistent with the requirements of special relativity. This quantum state is four- dimensional, with the first two components corresponding to the particle (e.g. the electron) and the second two components corresponding to the anti-particle (e.g. the positron). As a consequence, these Dirac spinor matrices γ^μ ($\mu = 0, 1, 2, 3$) are themselves 4×4 matrices. They satisfy the following 'Clifford algebra' relationship;

$$\{\gamma^\mu, \gamma^\nu\} = \gamma^\mu\gamma^\nu + \gamma^\nu\gamma^\mu = 2\eta^{\mu\nu} I_{4\times4}$$

where I_{4x4} is the 4×4 identity matrix (1s down the diagonal and 0 elsewhere) and $\eta^{\mu\nu}$ is the metric tensor of flat space time. From this relationship it follows that $\gamma^{\mu 2} = \eta^{\mu\mu}$.

In essence, as explained in detail in chapter 1, by finding a set of matrices which satisfied this condition, Dirac was able to take the square root of the relativistic relationship $E^2 = p^2c^2 + m^2c^4$ with p the momentum of the particle. When put into quantum operator terms, this results in both positive and negative square roots, corresponding to the particle and its anti-particle. These Dirac spinor matrices automatically incorporate the spin of the particle, showing that spin is an inherently relativistic effect.

In our representation we define $\frac{1}{2}\Sigma^{\mu\nu} = \frac{i}{4}[\gamma^{\mu}, \gamma^{\nu}]$ using the commutator of the Dirac matrices. From the properties of these matrices it can then be shown, by direct calculation, that the $\Sigma^{\mu\nu}$ in essence satisfy all the same algebraic relationships as the $M^{\mu\nu}$ ($\mu, \nu = 0, 1, 2, 3$) and thus form an operator valued representation. The full quantum operator representation is of the following form;

$$P_{\mu} \to i\partial^{\mu} = i\frac{\partial}{\partial x^{\mu}}$$

$$M_{\mu\nu} \to ix^{\mu}\partial^{\nu} - ix^{\nu}\partial^{\mu} + \frac{1}{2}\Sigma^{\mu\nu}.$$

We take here the Weyl representation of the Dirac matrices as in chapter 1 for which the left handed Weyl spinor quantum state ψ_L corresponds to the particle (e.g. the electron) and the right handed Weyl spinor quantum state ψ_R corresponds to the anti-particle.

Let ψ be a general Dirac spinor, and a solution of the massless form of the Dirac equation;

$$i\gamma^{\mu}(\partial_{\mu} - ieA_{\mu})\psi = 0$$

where ψ has charge e and A_{μ} ($\mu = 0, 1, 2, 3$) is the vector potential corresponding to an external magnetic field. If we define the Dirac adjoint function $\overline{\psi} = \psi^{+}\gamma^{0}$ where $\psi^{+}=(\psi^{*})^{T}$ then taking complex conjugates of both sides of the Dirac equation above, followed by transposition yields the identity:

$$(\psi^{*})^{T}(-i(\gamma^{\mu *})^{T}(\partial_{\mu} + ieA_{\mu})) = 0.$$

We can tweak the Dirac spinor matrix γ^{0} by a Wick rotation so that $\gamma^{0}\gamma^{0} = I$ (the 4×4 identity matrix) leading to the equation:

$$-i\gamma^{\mu T}(\partial_{\mu} + ieA_{\mu})\overline{\psi}^{T} = 0.$$

The matrices $(-\gamma^{\mu T})$ also satisfy the Clifford algebra relations and there is a 4×4 non-singular matrix C such that $C^{-1}\gamma^{\mu}C = -\gamma^{\mu T}$. We can thus define the *charge conjugate spinor*

$$\psi^{c} = C\overline{\psi}^{T}.$$

It then follows that;

$$-i\gamma^{\mu}(\partial_{\mu} + ieA_{\mu})\psi^{c} = 0.$$

In other words, ψ^{c} is the corresponding solution of the Dirac equation with charge (+e) rather than (–e). Adding mass into the equation essentially just adds a constant term thus the same conclusion applies.

In the Weyl representation we can take;

$$C = i\gamma^{0}\gamma^{2} \text{ (for consistency with standard references)}$$

$$=i\begin{pmatrix} -\sigma^{2} & 0 \\ 0 & \sigma^{2} \end{pmatrix} \text{ where } \sigma^{2} \text{ is the second Pauli spin matrix } \begin{pmatrix} 0 & -i \\ i & 0 \end{pmatrix}.$$

With ψ the Dirac spinor $\begin{pmatrix}\varphi\\ \bar{\chi}\end{pmatrix} = \begin{pmatrix}\varphi_A\\ \bar{\chi}^{\dot{A}}\end{pmatrix}$ we have

$$\psi^{c} = C\bar{\psi}^{T} = C(\psi^{+}\gamma^{0})^{T} = i\gamma^{0}\gamma^{2}\gamma^{0T}\psi^{*} = i\begin{pmatrix}0 & -\sigma^{2}\\ \sigma^{2} & 0\end{pmatrix}\begin{pmatrix}\varphi^{*}\\ \bar{\chi}^{*}\end{pmatrix} = i\begin{pmatrix}-\sigma^{2}\bar{\chi}^{*}\\ \sigma^{2}\varphi^{*}\end{pmatrix}.$$

Now note that
$\varphi^{*}=(\varphi_A)^{*}=\bar{\varphi}_{\dot{A}}$ and $\chi = \chi^{A} = (\bar{\chi}^{\dot{A}})^{*}$.
Thus

$$\psi^{c} = i\begin{pmatrix}-\sigma^{2}\bar{\chi}^{*}\\ \sigma^{2}\varphi^{*}\end{pmatrix} = \begin{pmatrix}\begin{pmatrix}0 & -1\\ 1 & 0\end{pmatrix}\bar{\chi}^{*}\\ \begin{pmatrix}0 & 1\\ -1 & 0\end{pmatrix}\varphi^{*}\end{pmatrix} = \begin{pmatrix}\varepsilon_{AB}(\bar{\chi}^{\dot{A}})^{*}\\ \varepsilon^{\dot{A}\dot{B}}(\varphi_A)^{*}\end{pmatrix}$$

$$= \begin{pmatrix}\varepsilon_{AB}\chi^{A}\\ \varepsilon^{\dot{A}\dot{B}}\bar{\varphi}_{\dot{A}}\end{pmatrix} = \begin{pmatrix}\chi_B\\ \bar{\varphi}^{\dot{B}}\end{pmatrix} \in E = F \oplus \dot{F}^{*}$$

where the ε matrices are the 'spinor metric' 2×2 matrices defined earlier and used to raise and lower indices in the 2-spinor calculus.

A *Marjorana* spinor ψ_M is defined mathematically as a Dirac spinor such that $\psi = \psi^{c}$. In other words it is its own charge conjugate spinor. Then: $\begin{pmatrix}\varphi_A\\ \bar{\chi}^{\dot{A}}\end{pmatrix} = \begin{pmatrix}\chi_A\\ \bar{\varphi}^{\dot{A}}\end{pmatrix}$ hence the Marjorana Dirac spinor has the form $\psi_M = \begin{pmatrix}\varphi_A\\ \bar{\varphi}^{\dot{A}}\end{pmatrix}$ in terms of the 2-spin or φ.

Thus, given a 2-spinor φ_A we can construct a Marjorana 4-spinor as $\psi_M = \begin{pmatrix}\varphi_A\\ \bar{\varphi}^{\dot{A}}\end{pmatrix}$.

Finally we note that it is possible to construct 'Dirac covariant bilinear forms' as follows. Given two Dirac 4-spinors;

$$\psi = \begin{pmatrix}\varphi_A\\ \bar{\chi}^{\dot{A}}\end{pmatrix} \text{ and } \Phi = \begin{pmatrix}\rho_A\\ \bar{\eta}^{\dot{A}}\end{pmatrix}$$

we can construct bilinear forms such as:

$$\{\psi, \Phi\} \to \bar{\psi}\Phi = \psi^{+}\gamma_0\Phi = (\psi^{*})^{T}\gamma_0\Phi$$

$$= \begin{pmatrix}\bar{\varphi}_{\dot{A}}\\ \chi^{A}\end{pmatrix}^{T}\begin{pmatrix}0 & \sigma^{0}\\ \sigma^{0} & 0\end{pmatrix}\begin{pmatrix}\rho_A\\ \bar{\eta}^{\dot{A}}\end{pmatrix} = \bar{\varphi}_{\dot{A}}\bar{\eta}^{\dot{A}} + \chi^{A}\rho_A = \bar{\varphi}\bar{\eta} + \chi\rho.$$

2.9.1 The supersymmetry algebra

We start with the properties of a Z_2-graded Lie algebra $L = L_0 \oplus L_1$ where L_0 is the Poincaré algebra spanned by the generators P_μ, $M_{\mu\nu}$ $(\mu, \nu = 0, 1, 2, 3)$ and L_1 is spanned by the spinor charge generators Q_α. Each of these is an abstract quantum operator.

From the definition of a Z_2-graded Lie algebra L we have, for elements $X_\mu, X_\nu \in L$ with 'gradings' g_μ, g_ν; $(X_\mu, X_\nu) \to X_\mu \circ X_\nu = X_\mu X_\nu - (-1)^{g_\mu g_\nu} X_\nu X_\mu$.

Thus for elements

$P_\mu \in L_0$, $Q_\alpha \in L_1$ with gradings 0 and 1

$$(P_\mu, Q_\alpha) \to P_\mu \circ Q_\alpha = P_\mu Q_\alpha - (-1)^{0x1} Q_\alpha P_\mu = P_\mu Q_\alpha - Q_\alpha P_\mu = [P_\mu, Q_\alpha] = 0.$$

Similarly,

$$M_{\mu\nu} \circ Q_\alpha = [M_{\mu\nu}, Q_\alpha] = -(\sigma^4_{\mu\nu})_{\alpha\beta} Q_\beta$$

with $\sigma^4_{\mu\nu}$ a 4×4 matrix representation of the Poincaré group.

We also have $Q_\alpha \circ Q_\beta = Q_\alpha Q_\beta - (-1)^{1\times 1} Q_\beta Q_\alpha = Q_\alpha Q_\beta + Q_\beta Q_\alpha = \{Q_\alpha, Q_\beta\}$, the anticommutator.

Each abstract operator Q_α, in a suitable representation space F, corresponds to a Dirac 4-spinor. Since Q_α is a non-Hermitian operator (by construction), we can also consider the complex-conjugate Dirac 4-spinor $Q_\beta \in \dot{F}$. To resolve different notations in the literature we can also write $\bar{Q}_\beta = \bar{Q}_{\dot{\beta}}$.

To constrain the number of options it is convenient at this stage to assume that Q_α is a Marjorana 4-spinor, thus $Q = Q^c = C\bar{Q}^T$.

Thus $(CQ)^T = (C^2\bar{Q}^T)^T = -\bar{Q}$ since $C^2 = -1$.

Hence $Q^T C = \overline{Q}$ since $C^T = -C$.

Consider now from the Z_2-grading, $\{Q_\alpha, Q_\beta\} = Q_\alpha Q_\beta + Q_\beta Q_\alpha = a(\gamma^\mu C)_{\alpha\beta} P_\mu$.

Thus (multiplying from right by C), $Q_\alpha Q_\beta C_{\beta\delta} + Q_\beta C_{\beta\delta} Q_\alpha = a(\gamma^\mu)_{\alpha\gamma} C_{\gamma\beta} C_{\beta\delta} P_\mu = -a(\gamma^\mu)_{\alpha\delta} P_\mu$.

$$\text{Hence } Q_\alpha (Q^T C)_\delta + (Q^T C)_\delta Q_\alpha = -a(\gamma^\mu)_{\alpha\delta} P_\mu,$$
$$\text{thus} \{Q_\alpha, \overline{Q}_\beta\} = -a(\gamma^\mu)_{\alpha\delta} P_\mu$$

We are now in a good position to proceed to the next chapter. We will exploit this last equation and related results to show that the graviton is compatible with the Dirac spinor calculus and more broadly with our general approach to quantum gravity.

References

[1] Kadison R and Ringrose J 1983 *Fundamentals of the Theory of Operator Algebras* vols 1 and 2 (New York: Academic)

[2] Einstein A 1960 *The Meaning of Relativity* 6th edn reprinted (London: Methuen)

[3] Dirac P 1958 *The Principles of Quantum Mechanics* 4th edn (Oxford: Oxford University Press)

[4] Regge T 1961 General relativity without coordinates *Nuovo Cimento* **19** 558–71

[5] Moffat J, Oniga T and Wang C H-T 2017 Unitary representations of the translational group acting as local diffeomorphisms of space-time *J. Phys. Math.* **8** 2

[6] Moffat J, Oniga T and Wang C H-T 2017 A new approach to the quantisation of paths in space-time *J. Phys. Math.* **8** 2
[7] Moffat J, Oniga T and Wang C H-T 2017 Ergodic theory and the structure of non-commutative space-time *J. Phys. Math.* **8** 2
[8] Moffat J, Oniga T and Wang C H-T 2017 Factorial unitary representations of the translational group, invariant pure states and the supersymmetric graviton *J. Phys. Math.* **8** 257
[9] Moffat J and Wang C 2018 Factorial representations of compact Lie groups, Wigner sets and locally invariant quantum fields *J. Phys. Math.* **9** 1
[10] Moffat J and Wang C 2018 Towards a non-perturbative quantum gravity *J. Phys. Math.* **9** 289
[11] Rovelli C and Smolin L 1995 Discreteness of area and volume in quantum gravity *Nucl. Phys.* B **442** 593–619
[12] Sinclair A and Smith R 1995 *Hochschild Cohomology of von Neumann Algebras* (Cambridge: Cambridge University Press)
[13] Maulik D, Nekrasov N, Okounkov A and Pandharipande R 2006 Gromov–Witten theory and Donaldson–Thomas theory *Compos. Math.* **142** 1263–85
[14] Kontsevich M and Manin Y 1994 Gromov–Witten classes, quantum cohomology, and enumerative geometry *Commun. Math. Phys.* **164** 525–62
[15] Li A and Ruan Y 2001 Symplectic surgery and Gromov–Witten invariants of Calabi–Yau 3-folds *Invent. Math.* **145** 151–218
[16] Requardt M 2000 Space-time as an order-parameter manifold in random networks and the emergence of physical points *Quantum Theory and Symmetries (Goslar, 1999)* (River Edge, NJ: World Scientific)
[17] Smolin L 2005 *An Invitation to Loop Quantum Gravity* (arxiv [hep-th 0408048]).
[18] Moffat J 2010 *Complexity Theory and Network Centric Warfare* (Darby, PA: DIANE) Also available on Amazon Kindle™
[19] Eilenberg S and Steenrod N 2015 *Foundations of Algebraic Topology* (Princeton, NJ: Princeton University Press)
[20] Nunke R and Rotman J 1962 Singular cohomology groups *J. Lond. Math. Soc.* **37** 301–6
[21] Wald R 1994 *Quantum Field Theory in Curved Space-Time and Black Hole Thermodynamics* (Chicage, IL: University of Chicago Press)
[22] Greiner W and Reinhardt J 1996 *Field Quantization* (Berlin: Springer)
[23] Pontryagin L 1966 *Topological Groups* 2nd edn (London: Gordon & Breach)
[24] Gell-Mann M and Low F 1954 Quantum electrodynamics at small distances *Phys. Rev.* **95** 1300–12
[25] Goddard P (ed) 1998 *Paul Dirac; The Man and his Work* (Cambridge: Cambridge University Press)
[26] Dimock J 2011 *Quantum Mechanics and Quantum Field Theory* (Cambridge: Cambridge University Press)

Chapter 3

Particles in algebraic quantum gravity

If any beauty I did see, Which I desired and got, 'Twas but a dream of thee'[1]

3.1 Introduction

In this chapter, we begin by redeeming our promise to have a deeper dive into fibre bundle theory. Our application of these structures includes Lie groups acting as gauge groups of automorphisms on the individual fibre spaces, so we will start with a deep dive into Lie group theory. In general, we will restrict ourselves to those parts of these two subjects most relevant to application in quantum gravity. We then develop the mathematics of supersymmetry, the massless graviton and the massive bosons and fermions. We assume a broken form of supersymmetry compatible with the latest results from the LHC ATLAS detector, allowing exploitation of the mathematics of the heterotic string.

Notation convention: in the following paragraphs, the Einstein summation convention applies (i.e. sum over repeated pairs of indices).

3.2 Lie groups, fibre bundles and quantum fields in loop quantum gravity

A Lie group G is a set of elements g of the form $\{g \in G;\, g = M(\varphi^1, \varphi^2, \ldots\ldots\ldots, \varphi^n)\}$ where M is an n-dimensional smooth manifold, and which satisfy the axioms for a group. The values of the contravariant one-forms $\varphi^1, \varphi^2, \ldots\ldots\ldots, \varphi^n$ form a set of continuous coordinates which define each element g of the group uniquely. The coordinate origin (0,0,…,0) corresponds to the identity e of the group so that $M(0,0,\ldots,0) = e$.

[1] This fragment of verse, written by John Donne (1572–1631), was quoted by Steven Weinberg in his book *Dreams of a Final Theory*.

doi:10.1088/978-0-7503-2580-6ch3

Let φ_j^α be a typical coordinate element of the manifold, so that $M(\varphi_j) = (\varphi_j^1, \ldots\ldots, \varphi_j^n)$. Under the group multiplication we have $M(\varphi_1)M(\varphi_2) = M(f(\varphi_1, \varphi_2))$ since the group is closed under multiplication, where $f(\varphi_1, \varphi_2)$ is a function of the two elements φ_1, φ_2. We also have the following constraints. Setting $\varphi_2 = 0$ we have $M(\varphi_2) = e$ and thus;

$$M(\varphi_1) = M(f(\varphi_1, 0)).$$

This implies that $\varphi_1^\alpha = f^\alpha(\varphi_1, 0)$. Similarly $\varphi_2^\alpha = f^\alpha(0, \varphi_2)$. Finally, if $\varphi_1 = \varphi_2 = 0$ then we have $M(0) = e = ee = M(0)M(0) = M(f(0,0))$. Hence $0 = f^\alpha(0, 0)$ for all $\alpha = 1,2,\ldots n$.

The set $\{M(\varphi^1,\varphi^2,\ldots,\varphi^n)|$ each φ^α runs through a continuous set of values$\}$ is an example of a *monster group*. This monster is 'tamed' by considering only;

1. Those groups which are connected to the identity.
2. Those groups for which a property of a small neighbourhood of the identity can be extrapolated to the whole group.

These assumptions are applicable to the Poincaré group P and imply the following. P is a non-trivial group thus there is a generator element g in P not equal to the identity e. It follows that there is a continuous path of elements connecting g to e. If W is a neighbourhood of the identity, then the path must pass through W. If W is a measurable set then this implies that W has positive, non-zero measure, since it contains a non-trivial interval. This fact will be useful later on in application to the subgroup T described below, so please remember it!

These properties 1 and 2 also apply to the Lie subgroup T of the Poincaré group corresponding to the group of translations of space–time, introduced in chapter 2. Let φ be an infinitesimal of first order, so that φ^2 is negligible. Then continuity implies that $g = M(\varphi)$ is close to the origin $e = M(0)$.

Let X_α be the elements of the Lie algebra g which generate the group, so that;

$$M(\varphi) = \exp(\varphi^\alpha X_\alpha).$$

Consider now a representation U of T, the translational group, as operators t_α acting on a Hilbert space H. To first order we have;

$$U(g) = U(M(\varphi)) = I + i\varphi^\alpha t_\alpha.$$

$U(g)$ a unitary then implies that $(U(g) * U(g)) = I$ and thus, to first order;

$$I = (I - i\varphi^\alpha t^*_\alpha)(I + i\varphi^\alpha t_\alpha) = I - i\varphi^\alpha t^*_\alpha + i\varphi^\alpha t_\alpha = I - i\varphi^\alpha(t^*_\alpha - t_\alpha).$$

It then follows that $\varphi^\alpha(t^*_\alpha - t_\alpha) = 0$ for all α and all φ. Taking $\varphi_j^\alpha = 1$ for $\alpha = j$ and $=0$ for $\alpha \neq j$. We have $\varphi_1 = (1,0,0,\ldots\ldots,0)$; $\varphi_2 = (0,1,0,\ldots\ldots,0)$ etc. These then imply that;

$$t^*_\alpha = t_\alpha \text{ for all } \alpha = 1, 2, 3, \ldots, n.$$

Thus each of these t_α is a Hermitian linear operator, corresponding to an observable of the quantum system, acting on the Hilbert space H. Define the product of two such operators as $t_a \circ t_b = [t_a, t_b] = if^c_{ab}t_c$ with $f^c_{ab} = -f^c_{ba}$ so that f is antisymmetric. The values f^c_{ab} are the *structure constants* of the Lie group. It can then be shown [1] that the set $\{t_a;\ a = 1, 2, \ldots n\}$ is a representation of the Lie algebra of the group T.

A *fibre bundle* is a triple (E, π, B) where E is a topological space and π is a projection onto a subset B called the base space. For a point x in B, the set $F = \{y \in E \cap \pi^{-1}(x)\}$ is the *fibre* at the point x, and it is assumed that the bundle is locally a product space about each point x. E is a *vector bundle* if each fibre has the structure of a vector space. In our case we can go further and define the fibre at x as consisting of all bases of the vector space attached to the point x; and define the group G of changes of basis, acting on the fibre space F. This is an example of a *principal bundle*. A *section* or *lifting* s through the fibre bundle (E, π, B) is a continuous right inverse of the function π. In other words, s is a continuous function mapping from B to E such that $\pi(s(x)) = x$ for all x in B. In the approach due to Connes [2], commutative space–time is replaced at high energy by a principal product bundle $X = M \times F$ with M Minkowski space–time and the fibre F a finite vector space. Commuting functions become an algebra of non-commuting sections through this bundle. These sections are liftings from flat space–time M into X in the sense that a point $x \in M$ is lifted into the fibre rooted at x. Diffeomorphisms between reference frames become (in the Schrödinger picture) unitary transformations U between different bases on F, and thus (in the Heisenberg picture) operator automorphisms of the algebra of sections. These are of the form $A(x) \rightarrow UA(x)U^*$ for a point x of space–time and a section function $s: x \rightarrow A(x)$ for all x in M.

3.2.1 A coherent approach to quantum fields

In terms of a coherent quantum field theory, the implications are as follows:

(a) Each local algebra $O(D)$ should correspond to the algebra of sections of a principal fibre bundle with base space a finite and bounded subset of space–time as discussed by Araki [3]. The algebraic operations of addition and multiplication are assumed defined fibrewise for the algebra of sections.

(b) The group of unitaries $U(x)$ acting on the fibre space $F(x) = \pi^{-1}(x)$ corresponds to a local gauge group. The fibre space is assumed to be a separable Hilbert space of finite or at most countable dimension.

(c) There is, as shown in chapter 2, a limit on the resolution of the space–time variables and a corresponding limit to the energy available to the system—i.e. an upper energy or 'UV cut-off'. This is reflected in the fact that locally the fibre bundle is a product bundle, and that neighbouring fibres share the same locally flat space–time.

Unitary representations corresponding to local shifts in reference frames are a key topic for progress in loop quantum gravity. Our proposed construct is such that a quantum field becomes a section through a fibre bundle with base space an emergent compact subset of curved space–time, evolved from a quantitative spin network. We investigate in more depth the subgroup T of the Poincaré group consisting of translations of space–time as a gauge group of automorphisms. We define a representation of T as a gauge group of automorphisms of the local fibre algebra **A(x)** which we assume to be isomorphic to a von Neumann algebra with trivial centre acting on a separable Hilbert space. These automorphisms are the quantum generators equivalent to the local diffeomorphisms of general relativity.

Consider then the automorphic representation $g \to \alpha_g$ of the translation subgroup T. If we make the minimal assumption that this representation is weakly measurable; i.e. the mapping $g \to \langle \alpha_g(A)x, y \rangle$ is Haar-measurable for all relevant values of A, x and y; then the argument of [4] shows that the mapping $g \to \alpha_g$ is norm continuous; $\|\alpha_g - i\| \to 0$ as $g \to e$. This then leads to the following remarkable theorem which I proved, in essence, in 1973 and published [5] when doing my PhD. Details are given in the Preface to this book of how to access a scanned form of my PhD thesis. I had to first laboriously write out the thesis by hand and then it was typed—hence the rather hand-crafted appearance of the scanned text.

3.3 A remarkable theorem

Let **A** be a von Neumann algebra acting on a general Hilbert space H which is not necessarily separable. Let $g \to \alpha_g$ be an automorphic representation of the translation subgroup T acting on **A**. We make the minimal assumption that this representation is weakly measurable; i.e. the mapping $g \to \langle \alpha_g(A)x, y \rangle$ is Haar-measurable for all relevant values of A, x and y.

It then follows that;

- The mapping $g \to \alpha_g$ is norm continuous; $\|\alpha_g - i\| \to 0$ as $g \to e$.
- Each α_g is an inner automorphism of **A** and the corresponding unitary W_g has a spectrum contained in the positive half plane.
- The set of unitaries $\boldsymbol{S} = \{W_g | g \in G\}$ can be replaced by a unitary representation $g \to U_g$ of the translational group T acting on the Hilbert space H which is norm continuous hence strongly and weakly continuous.
- The unitary representation $g \to U_g$ *implements* the automorphic representation $g \to \alpha_g$. This means that for all elements A of the algebra **A** we have $\alpha_g(A) = U_g A U_g^*$.

The proof falls into two parts. In the first part we prove that a Haar-measurable automorphic representation $g \to \alpha_g$ of the translational group as an *l.c.a.* (locally compact abelian) group is norm continuous. We do this first part by exploiting the result for a weaker form of continuity which I proved in [6], also in 1973, clearly a good year for my research ideas. The essence of this part of the proof centres on the fact that if W is the non-trivial measurable set

$$W = \{g \in T | \|\alpha_g - i\| \leqslant \frac{\varepsilon}{2}$$

containing the identity e of T (since $\alpha_e = i$); then

- $W = W^{-1}$

for we have for $h \in W^{-1}$; since each automorphism is an isometry,

$$\|\alpha(h^{-1}) - i\| = \|\alpha(h^{-1})i - \alpha(h^{-1})\alpha(h)\| \leqslant \|i - \alpha(h)\| = \|\alpha(h) - i\|.$$

Thus $W^{-1} \subseteq W$. The equality of these two sets then follows by replacing h by h^{-1} in the above discussion.

- $W^2 \subseteq \{g \in T \| \|\alpha_g - i\| \leqslant \varepsilon\}$ follows from the calculation, for g, h elements of the subgroup T;

$$\begin{aligned}\|\alpha(gh) - i\| &= \|\alpha(g)\alpha(h) - \alpha(g)\alpha(h)\alpha(h^{-1})\alpha(g^{-1})\| \\ &\leqslant \|\alpha(h) - \alpha(h)\alpha(h^{-1})\alpha(g^{-1})\| \\ &= \|\alpha(h) - \alpha(hh^{-1}g^{-1})\| \\ &= \|\alpha(h) - \alpha(g^{-1})\| \leqslant \|\alpha(h) - i\| + \|i - \alpha(g^{-1})\| \leqslant \varepsilon.\end{aligned}$$

It follows from [7] that there is a neighbourhood N of the identity e of T, and a compact set C with $N \subseteq CC^{-1} \subseteq W^2 \subseteq \{g \in T \| \|\alpha_g - i\| \leqslant \varepsilon\}$. Thus the representation α is norm continuous.

The second part of the proof centres around the existence of a lifting from the carrier space of the centre $\mathcal{Z}$ of **A** into the von Neumann algebra generated by the set $\mathcal{Z} \cup S$. We continue to label this von Neumann algebra as **S**. Such a commutative quantum operator algebra is equivalent to the set of continuous functions on a compact space and this equivalence arises through the Gelfand transform; $A \to \hat{A}$ with $\hat{A}(\rho) = \rho(A)$ for ρ a continuous complex valued homomorphism.

The carrier space Φ_S of the von Neumann algebra **S** is the set of all continuous complex valued homomorphisms on **S** and is thus topologically a Stonean space, as is Φ_Z and the restriction map $\pi\colon \Phi_S \to \Phi_Z$ is a continuous surjection. We then have, as we will prove, a lifting

$$f\colon \Phi_Z \to \Phi_S \text{ with } f(\rho)|_Z = \pi \circ f(\rho) = \rho \text{ for } \rho \in \Phi_Z.$$

For each $g \in T$, we can now define the Gelfand transform of a unitary U_g acting on the carrier space Φ_S of **S** as follows;

$$\hat{U}_g(\rho) = \bar{\hat{W}}_g(f(\rho|_Z))\hat{W}_g(\rho), \quad \rho \in \Phi_S.$$

If we define the equivalence $\rho \approx \rho' \Leftrightarrow \rho|_Z = \rho'|_Z$ then by the extended form of the Stone–Weierstrass theorem [8] the centre **Z** corresponds to those elements of Φ_S constant on each equivalence class. Applying this to the Gelfand transform $\hat{V}_g(\rho) = \bar{\hat{W}}_g(f(\rho|_Z))$ it follows that $\hat{V}_g$ corresponds to an element V_g of the centre. This implies that for each $g \in T$; U_g implements α_g. In addition, the mapping $g \to U_g$ is a group homomorphism for if we set;

$$R_{g,h} = U_g U_h U_{gh}^*$$

then for an operator

$$\begin{aligned}A,\ R_{g,h} A R_{g,h}^* &= U_g U_h U_{gh}^* A U_{gh} U_h^* U_g^* \\ &= \alpha_g \alpha_h \alpha_{gh}^{-1}(A) = A.\end{aligned}$$

Hence, $R_{g,h}$ is a unitary in the centre.

The fact that the lifting f: $\Phi_Z \to \Phi_S$ has the property $\pi \circ f(\rho) = f(\rho)|_Z = \rho$ for $\rho \in \Phi_Z$ means that $\hat{R}_{g,h}(\rho) = \hat{R}_{g,h}(f(\rho))$ since the domain of $\hat{R}_{g,h}$ is Φ_Z. But then we have that $\hat{R}_{g,h}(\rho) = 1 \circ \ \forall \rho \in \Phi_Z$. Therefore $R_{g,h} = I \circ \ \forall g, h \in G$ and $g \to U_g$ is a group representation by unitaries in **A** which turns out to be norm continuous, as previously proved in reference [5]. We note in passing that this theorem probably extends to the whole of the Poincaré group P by considering in turn each of its abelian subgroups. We leave this as an exercise for you, dear reader.

3.4 Properties of the projection onto the base space *B* of a Stonean fibre bundle *K*

We prove first the subsidiary fact that the projection π is an open mapping if and only if for each non-trivial open subset E of K, $\pi(E)$ is not a nowhere dense subset of the base space B. One way is trivial for if π is an open mapping then $\pi(E)$ is a non-trivial open set so cannot be nowhere dense.

Conversely, the topology of the fibre bundle K is compact and totally disconnected, with a basis of *clopen* sets (i.e. sets which are both closed and open). Thus every open set is a union of such clopen sets, and it suffices to show that if V is clopen in K, then $\pi(V)$ is open in B. Since V is clopen, it is a closed and thus compact subset of K, and π is continuous, thus $\pi\ (V)$ is compact. If we define $Y = \pi(V) \backslash \operatorname{int} \pi(V)$; this is a closed set with empty interior thus Y is a nowhere dense set, and is the image of an open set; $Y = \pi(V \backslash \pi^{-1}(\operatorname{int} \pi(V))$, using the fact that the projection mapping π is continuous and surjective. It follows that Y is void and $\pi(V) = \operatorname{int} \pi(V)$. Thus $\pi(V)$ is open and π is an open mapping.

3.4.1 Lifting from the base space

Consider now the set Φ of all compact subsets S of K such that $\pi(S) = B$. Then Φ is non-void, partially ordered by set inclusion, and every decreasing chain has a lower bound. It follows from Zorn's Lemma that Φ has a smallest element $K(0)$. We show that;

- $\pi|_{K(0)}$ is an open mapping;
- $\pi|_{K(0)}$ is injective.

This will prove the result; for the lifting we seek is then $f = (\pi|_{K(0)})^{-1}$.

The axiom of choice selects out $K(0)$ as a unique minimal section through the fibre bundle K corresponding to a quantum field.

$\pi|_{K(0)}$ is an open mapping.

Consider then V to be a non-trivial open set in $K(0)$. By definition, $\pi|_{K(0)}$ is surjective, thus; $B \backslash \pi(V) \subset \pi(K(0) \backslash V)$. Now $K(0) \backslash V$ is a closed thus compact subset of K, and $\pi|_{K(0)}$ is continuous, thus $\pi(K(0) \backslash V)$ is compact. It follows that $(B \backslash \pi(V))^{-} \subset \pi(K(0) \backslash V)$. If $\pi(V)$ is nowhere dense, then;

$$B = B\backslash \text{ int } (\pi(V))^{-} \subset B\backslash\pi(V)^{-} \subset \pi(K(0)\backslash V),$$

a closed, compact set.

This is a contradiction due to minimality of the set $K(0)$. Thus $\pi(V)$ cannot be a nowhere dense set. From our earlier discussion, this shows $\pi|_{K(0)}$ is an open mapping.

$\pi|_{K(0)}$ is injective.

Assume that $\pi|_{K(0)}$ is not injective, then $\exists\ x_1, x_2;\ x_1 \neq x_2 \in K(0)$ with $\pi(x_1) = \pi(x_2)$.

Since $K(0)$ is a Hausdorff topological space with a basis of clopen sets, there is a clopen subset V of $K(0)$ containing x_1 but not x_2. Then, $\pi(V)$ is also clopen and $\pi^{-1}(\pi(V))\backslash V \subset K(0)$, and $\pi(x_2) = \pi(x_1) \in \pi(V)$, thus $x_2 \in \pi^{-1}(\pi(V))\backslash V$. It follows that $W = \pi^{-1}(\pi(V))\backslash V$ is a non-trivial open set, and W and V are disjoint, with $B\backslash\pi(V) \subset \pi(K(0)\backslash W)$ and $\pi(V) \subset \pi(K(0)\backslash W) \Rightarrow \pi(K(0)\backslash W) = B$.

This again contradicts the minimality of $K(0)$. Thus $\pi|_{K(0)}$ is a continuous bijection, and the required lifting is given by $f = (\pi|_{K(0)})^{-1}$.

3.5 Quantum connections

We now define a connection on the bundle $O(D)$ of Ehresmann type [9], tailored to our requirements.

Definition. If π is the projection map from $O(D) \rightarrow D$, let $\mathbf{x}$ be an element of D and p an element of the fibre $\pi^{-1}(\mathbf{x})$, so that $\pi(p) = \mathbf{x}$. The 'push forward' of π, denoted π_* is a connection we define as follows. Let $t \rightarrow A(x(t))$ be a section passing through the point p in the fibre $\pi^{-1}(\mathbf{x})$, so that $p = A(x(t_0))$, and $\pi(p) = x(t_0)$, a point on the curve $t \rightarrow \mathbf{x}(t) = \pi \circ A(x(t))$ defined on the base space D and passing through the point $\mathbf{x}(t_0)$ with velocity $v_{\mathbf{x}} = \frac{\partial \mathbf{x}}{\partial t}|_{t_0}$. If $v_p = \frac{\partial(Ax(t))}{\partial t}|_{t_0}$, with v_p the velocity of the curve $t \rightarrow A(x(t))$ at p, then $\pi_*(v_p) = v_{\mathbf{x}}$.

We now define the tangent space to $O(D)$ as the linear space $TO(D)$ generated by the set

$$\left\{\frac{\partial A(x(t))}{\partial x_\mu};\ t \rightarrow A(x(t)) \text{ is a path in } O(D) \text{ and } x_\mu = t, x, y, z\right\}$$

Similarly, we define the tangent space to D, denoted TD, as the linear space generated by the set;

$$\left\{\frac{\partial(x(t))}{\partial x_\mu};\ t \rightarrow x(t) = \mathbf{x}(t) \text{ is a path in } D \text{ and } x_\mu = t, x, y, z\right\}$$

With these definitions, we see that the connection π_* maps $TO(D)$ to TD.

Note that if the basis of $TO(D)$ is defined by Clifford variables e_μ, then $TO(D)$ contains the range of the Dirac operator $D = \sum_\mu e_\mu \frac{\partial}{\partial x_\mu}$.

A *quantum geodesic* is a path in $O(D)$ between operators $A(x(0))$ and $A'(x(t_0))$ whose projection onto the base space D is a geodesic connecting the space–time points $x(0)$ and $x(t_0)$; defined in terms of the Christoffel tensor of the gravitational field.

From the Heisenberg equation of motion we have that;

$$v_p = \frac{\partial(Ax(t))}{\partial t} = \frac{1}{i\hbar}[Ax(t), H]$$

and from classical mechanics we have, from chapter 2, the use of a 2-form $\Omega^{\mu\nu}$ which essentially imposes a symplectic structure on phase space. This is of the form of a symplectic *vector bundle* $V \oplus V^*$ where V is a real finite vector space with dual V^*. The skew-symmetric rank 2 tensor Ω then takes the general form;

$$\Omega(x \oplus \eta, x' \oplus \eta') = \eta' \cdot x - \eta \cdot x'.$$

3.6 The supersymmetric extension of the standard model

The basic idea of supersymmetry is to add additional vector elements, normally denoted $Q_{\alpha j}$; $j = 1, \ldots, N$, to the set of abstract tensors and vectors $\{M_{\mu\nu}, P_\mu\}$ which form the Lie algebra of the Poincaré group. For the case $N = 1$, we need to add one additional element to the Lie algebra of the Poincaré group. In this case we have the minimal supersymmetric extension of the standard model (MSSM). We will focus on this $N = 1$ form as being most compatible with a four-dimensional string solution. In any representation, these are all linear operators, including the identity operator I, and form part of an algebra of such operators. This is done in such a way that an irreducible representation corresponds not to a single particle but to a set of particles, called a *multiplet*, all of the same mass m, (if the symmetry is not broken) and containing equal numbers of bosons (particles with integer spin values such as 0 or 1) and fermions (particles with half integer spin such as 1/2).

The set $\{M_{\mu\nu}, P_\mu, Q_{\alpha j}\}$ no longer form a standard Lie algebra, but in fact form a graded Lie algebra. As discussed in some detail in chapter 2, this means that additional mathematical machinery is required to define multiplication in this enhanced algebra, resulting in the ability to mix together both commutators and anticommutators. Salaam and Strathdee [10] take this further by developing a non-standard Lie *superalgebra*. In any representation of a superalgebra, the elements are all linear operators, including the identity operator I, and form an algebra of such operators.

In chapter 2 we also discussed the reasoning leading to the conclusion that an irreducible representation π of the Poincaré Lie algebra implies that Casimir operators P^2 and W^2, in the centre of our algebra take the form $\pi(P^2) = \lambda_1 I$; $\pi(W^2) = \lambda_2 I$. Our black box assumptions mean that our mathematical representation produces the values λ_1, λ_2 and these values correspond to a single particle with mass m and spin value s. Adding the terms $Q_{\alpha j}$ to form our superalgebra implies that the operator W^2 can no longer commute with this larger set of operators. We now only have the single Casimir operator P^2. In an irreducible representation of our superalgebra, this constrains the mass, However, the spin values are no longer constrained except that such an irreducible representation must contain equal numbers of bosons and fermions.

In summary, an irreducible representation of the Poincaré–Lie superalgebra corresponds to a *multiplet* of particles of the same mass but with differing spin values; there are equal numbers of particles with integer spin (bosons) and half-integer spin (fermions). This theoretical symmetry between bosons and fermions is called *supersymmetry*.

We can develop a locally linear, faithful representation of the superspace Lie algebra by adding a pair of Grassmann variables to the algebraic formulation. We can then generate a standard Lie algebra while mixing commutators and anti-commutators. For example, if ξ and $\bar{\xi}$ have the Grassmann property so that $\xi\bar{\xi} = -\bar{\xi}\xi$, then;

$$[\xi Q_A, \bar{\xi}\bar{Q}_{\dot{B}}] = (\xi Q_A\bar{\xi}\bar{Q}_{\dot{B}} - \bar{\xi}\bar{Q}_{\dot{B}}\xi Q_A) = \xi\bar{\xi}(Q_A\bar{Q}_{\dot{B}} + \bar{Q}_{\dot{B}}Q_A) = \xi\bar{\xi}\{Q_A, \bar{Q}_{\dot{B}}\}.$$

Note that we also require that these Grassmann variables commute with the operators Q. A typical unitary element of the corresponding Lie group G is then of the form;

$$G(x^\mu, \xi, \bar{\xi}) = \exp i(\xi Q + \bar{\xi}\bar{Q} - x^\mu P_\mu).$$

Here, x^μ is an event point in locally flat space–time, thus we can think of the Grassmann variables as a vector at the point x^μ. This structure is similar to that envisaged by Connes [2]. The superspace is a vector bundle and the locally flat group multiplication structure is of the following form;

$$G(x^\mu, \theta, \bar{\theta})G(a^\nu, \xi, \bar{\xi}) = G(x^\mu + a^\nu - i\xi\sigma^\mu\bar{\theta} + i\theta\sigma^\mu\bar{\xi}, \theta + \xi, \bar{\theta} + \bar{\xi}).$$

This follows from the Grassmannian properties, since $\theta^2 = \bar{\theta}^2 = \xi^2 = \bar{\xi}^2 = 0$.

We can interpret this group product as shifting the locus of the Grassmann vector in space–time from x^μ to $x^\mu + a^\mu - i\xi\sigma^\mu\bar{\theta} + i\theta\sigma^\mu\bar{\xi}$ together with additive change to the Grassmann vector at this point. If this shift is infinitesimal, then, as in normal Lie group theory, we can consider the tangent plane around the group element $G(a^\mu, \xi, \bar{\xi})$, giving the following local representation of the Lie group generators on the tangent plane to the emergent Riemannian space–time manifold at the event point (a^ν):

$$P_\mu(a^\nu) = i\frac{\partial}{\partial x^\mu}|_{(a^\nu)} = i\partial_\mu|_{(a^\nu)}$$

$$iQ_A(a^\nu) = \frac{\partial}{\partial\theta_A}|_{(a^\nu)} - i\sigma^\mu\bar{\theta}\frac{\partial}{\partial x^\mu}|_{(a^\nu)}$$

$$i\bar{Q}_{\dot{A}}(a^\nu) = -\frac{\partial}{\partial\bar{\theta}_{\dot{A}}}|_{(a^\nu)} + i\theta\sigma^\mu\frac{\partial}{\partial x^\mu}|_{(a^\nu)}.$$

In this form the generators satisfy all the algebraic relationships of the super-symmetric extension of the local translational Lie algebra. Thus, this representation is locally an algebraic isomorphism onto the curved Riemannian manifold M and the piecewise local representations;

$$P_\mu(a^\nu) = i\frac{\partial}{\partial x^\mu}|_{(a^\nu)} = i\partial_\mu|_{(a^\nu)}$$

are the generators of the relativistic diffeomorphisms around the event points of M. The manifold is assumed smoothly differentiable; we focus on the flat tangent plane and assume Dirac relativistic spinor theory applies. We therefore assume that this operator representation acts on a four-dimensional Hilbert space H of (spinor) wave functions; and we denote by ψ an element of the Hilbert space. The detailed calculations in chapter 2 then show that we have

$$\{Q_\alpha, \bar{Q}_\beta\} = -a(\gamma^\mu)_{\alpha\beta}P_\mu$$

where Q, $\bar{Q}$ are Marjorana 4-spinors. In 2-spinor notation we replace the Dirac γ matrices with their equivalent Pauli matrices yielding the following relationship;

$$\{Q_A, \bar{Q}_{\dot{B}}\} = -a(\sigma^\mu)_{A\dot{B}}P_\mu. \tag{3.1}$$

Similarly, we can show that, for 4-spinors,

$$\{\bar{Q}_\alpha, \bar{Q}_\beta\} = -a(C^{-1}\gamma^\mu)_{\alpha\beta}P_\mu.$$

Hence in 2-spinor notation,

$$\{\bar{Q}_A, \bar{Q}_{\dot{B}}\} = -a(C^{-1}\sigma^\mu)_{A\dot{B}}P_\mu. \tag{3.2}$$

Since all these latter equations are relativistically invariant, we can transform them to the rest frame where $P_\mu = (E, 0,0,0) = (m, 0,0,0)$ with m the rest mass and the speed of light $c = 1$.

With these values of P_μ in equation (3.1) above we have

$$\{Q_A, \bar{Q}_{\dot{B}}\} = -a\sigma^0_{A\dot{B}}P_0.$$

Hence

$$\{Q_A, \bar{Q}_{\dot{B}}\}\sigma^0_{\dot{B}A} = -a\sigma^0_{A\dot{B}}\sigma^0_{\dot{B}A}P_0 = -aP_0.$$

Taking $A = 1, \dot{B} = \dot{1}$ implies $\{Q_1, \bar{Q}_{\dot{1}}\}\sigma^0_{\dot{1}1} = -aP_0$; $A = 2, \dot{B} = \dot{2}$ implies $\{Q_2, \bar{Q}_{\dot{2}}\}\sigma^0_{\dot{2}2} = -aP_0$.

Since $\sigma^0_{\dot{1}1} = \sigma^0_{\dot{2}2} = 1$ we have;

$$\{Q_1, \bar{Q}_{\dot{1}}\} + \{Q_2, \bar{Q}_{\dot{2}}\} = -2aP_0.$$

For consistency with the current literature, we assume the constant $a = -1$.

Thus, we have the quantum operator equality

$$\{Q_1, \bar{Q}_{\dot{1}}\} + \{Q_2, \bar{Q}_{\dot{2}}\} = 2P_0. \tag{3.3}$$

The left-hand side of this expression is a positive definite quantum operator.

Let ψ be the vacuum state element of the Hilbert space, then;

$$\langle \psi, P_0 \psi \rangle = 0$$

From equation (3.3) this is equivalent to;

$$\langle \psi, Q_1 \bar{Q}_1 \psi \rangle + \langle \psi, \bar{Q}_1 Q_1 \psi \rangle + \langle \psi, Q_2 \bar{Q}_2 \psi \rangle + \langle \psi, \bar{Q}_2 Q_2 \psi \rangle = 0.$$

From chapter 2 we know that for 2-spinors; $\varphi_A^* = \bar{\varphi}_{\dot{A}}$ thus we can rewrite this as

$$\langle \psi, Q_1 Q_1^* \psi \rangle + \langle \psi, Q_1^* Q_1 \psi \rangle + \langle \psi, Q_2 Q_2^* \psi \rangle + \langle \psi, Q_2^* Q_2 \psi \geqslant 0.$$

Thus we have;

$$0 \leqslant 2||Q_1 \psi||^2 + 2||Q_2 \psi||^2 = 0.$$

We deduce that;

$$\omega_\psi(Q_1) = \omega_\psi(Q_2) = \omega_\psi\left(Q_1^*\right) = \omega_\psi\left(Q_2^*\right) = \omega_\psi(P_0) = 0.$$

3.7 Factorial representations of the graded Lie algebra

The extension of the space L_0 to the space $L = L_0 \oplus L_1$ maintains P^2 as an element of the centre;

$$[P^2, Q_A] = 0 = [P^2, \bar{Q}_{\dot{A}}].$$

In a factorial representation Π of the Z_2 graded algebra, it follows that $\Pi(P^2) = m^2 I$, fixing the mass m of all particles in this representation. However, as we discussed earlier, the spins of the particles in this representation are not fixed at a common value.

In this factorial 2-spinor representation, we have the algebraic identity; $\{Q_A, \bar{Q}_{\dot{B}}\} = \sigma^{\mu}_{A\dot{B}} P_\mu$ from equation (3.1) with the parameter $a = -1$.

If $|p, \lambda\rangle$ is an eigenstate in the Hilbert space, then $P_\mu |p, \lambda\rangle = p_\mu |p, \lambda\rangle$.

The corresponding (pure) vector state is invariant under the translational group since we have

$$0 = P_\mu Q_A |p, \lambda\rangle = Q_A P_\mu |p, \lambda\rangle = Q_A p_\mu |p, \lambda\rangle = p_\mu Q_A |p, \lambda\rangle.$$

We can thus always choose a minimum energy pure vector state $\omega_{|p,\lambda\rangle}$ which is translation invariant and with $\omega_{|p,\lambda\rangle}(Q_A) = \langle p, \lambda | Q_A | p, \lambda\rangle = 0$. It is specified by its mass–energy p and its spin value λ. Thus $\omega_{|p,\lambda\rangle}$ is a translation invariant ergodic pure state.

Exploiting local special relativistic covariance, we now choose an inertial reference frame in which the Wigner little group contains the spin generating 2×2 rotation matrices in the x–y plane. We have;

$$P_\mu = (E, 0, 0, E)$$

$$\{Q_A, \bar{Q}_{\dot{B}}\} |p, \lambda\rangle = \sigma^{\mu}_{A\dot{B}} P_\mu |p, \lambda\rangle = \sigma^{\mu}_{A\dot{B}} p_\mu |p, \lambda\rangle.$$

Now, for the 'mostly minus' case,

$$\sigma^{\mu}_{A\dot{B}}p_{\mu} = \sigma^0 p_0 - \sigma^1 p_1 - \sigma^2 p_2 - \sigma^3 p_3 = E(\sigma^0 - \sigma^3)$$
$$= E\left(\begin{pmatrix} 1 & 0 \\ 0 & 1 \end{pmatrix} - \begin{pmatrix} 1 & 0 \\ 0 & -1 \end{pmatrix}\right) = \begin{pmatrix} 0 & 0 \\ 0 & 2E \end{pmatrix}.$$

Hence applying this to our translation invariant vacuum state, $\omega_{|p,2\rangle}$ we have;

$$|\langle p,\, 2|Q_1\bar{Q}_{\dot{1}}|p,\, 2\rangle = -\langle p,\, 2|\bar{Q}_{\dot{1}}Q_1|p,\, 2\rangle = 0.$$

Similarly, considering the $2\dot{1}$ element of the matrix, we have;

$$|\langle p,\, 2|Q_2\bar{Q}_{\dot{1}}|p,\, 2\rangle = -\langle p,\, 2|\bar{Q}_{\dot{1}}Q_2|p,\, 2\rangle = 0.$$

Thus:

$$(aQ_1 + bQ_2)\bar{Q}_{\dot{1}}|p,\, 2\rangle = 0$$

for any scalars a and b.

Since Q_1 and Q_2 span the subspace L_1, it follows that $\bar{Q}_{\dot{1}}|p,\, 2\rangle = 0$.

We conclude that the translation invariant pure Boson state $\omega_{|p,2\rangle}$ is the local Clifford vacuum, and a spin-2 pure state corresponding to the massless graviton. It is an extreme point of the closed convex hull of the state space and is thus an extreme point of the translation invariant states: it is an ergodic state. Its supersymmetric fermion partner is the massless gravitino with spin $\frac{3}{2}$.

We conclude that:

- From the properties of factorial representations of the translational group which are the generators of local diffeomorphisms and gravity, a form of $N = 1$ supersymmetry naturally emerges, which is consistent with our algebraic approach.
- This theory is based on an extension to the Poincaré–Lie algebra which reduces the number of Casimir operators and predicts a multiplet consisting of a massless spin—2 graviton and its supersymmetric partner, a massless fermion, the gravitino, with spin $\frac{3}{2}$.
- Clearly, if we do not extend the Poincaré–Lie algebra then W remains a Casimir operator and we predict simply a spin—2 graviton with no superpartner.

Our mathematical approach as it has developed so far in chapters 1 and 2 shows that space–time is discrete during the Planck regime and thus finite in the sense of Dirac. Adding onto this the assumption of supersymmetry allows us to construct a coherent mathematical theory of the supersymmetric graviton which goes beyond the standard model. However, experiments designed to detect supersymmetry have given negative results so far. We summarise the experimental evidence below. This is based on aggregating data from the ATLAS detector at the upgraded Large Hadron Collider (LHC) running at 13 TeV over the period from 2015 to 2016. The summary that follows includes extracts from a journal paper [11] produced by the ATLAS collaboration in 2018 and reproduced by permission of the creative commons agreement therein.

3.8 Does supersymmetry exist? ATLAS results for Run 2 of the LHC at 13 TeV

As we have discussed, supersymmetry is a generalization of symmetries related to the Lie algebra of the Poincaré group which predicts new bosonic (resp. fermionic) partners for the fermions (resp. bosons) of the standard model. The lightest supersymmetric particle (LSP) is potentially stable, and represents a possible dark-matter candidate. Theoretical superpartners of the charged and neutral electroweak and Higgs bosons produce charginos and neutralinos, collectively known as electroweakinos. The production cross sections of such *sparticles* at the LHC depend both on the type of interaction involved and on the sparticle masses.

From the CERN Courier July 2015; 'LHC and Planck: where two ends meet'; we have the following optimistic assessment following the upgrade to the LHC [my comments];

'The main outcome of Run 1 [of the LHC] was undoubtedly the widely anticipated discovery of a Higgs boson [completing the Standard Model]. Run 2 [using the upgraded LHC at 13 TeV] is opening the door to uncharted territory. In practical and experimental terms, exploring the properties and the behaviour of nature at high energy consists in understanding possible signals that include 'missing energy'. In the Standard Model, this energy discrepancy is associated with neutrinos, but in physics beyond the Standard Model, the missing energy could also be the signature of many undiscovered particles, including the weakly interacting massive particles (WIMPs) that are among the leading candidates for dark matter. If WIMPs exist, the LHC's collisions at 13 TeV may reveal them, and this will be another huge breakthrough. Because supersymmetry has not yet been ruled out, the high-energy collisions might also eventually unveil the supersymmetric partners of the known particles, at least the lighter ones. Missing energy could also account for the escape of a graviton into extra dimensions, or a variety of other possibilities. Thanks to the LHC's Run 1 and other recent studies, the Standard Model is so well known that future observation of an unknown source of missing energy could be confidently linked to new physics ....'

The results of a recent review by the ATLAS collaboration of relevant data collected during Run 2 show no solid evidence for supersymmetry. These data were based on proton–proton collisions at 13 TeV in the LHC at the ATLAS detector over the period 2015 to 2016. The data were aggregated and analysed in 2018 to search for signals at the 5-standard deviation level of electroweak production of sparticles. This had to be done indirectly through backtracking from the final decay products; a process for which the ATLAS detector is ideally suited. This analysis was performed for a set of two-lepton and three-lepton final states using recursive jigsaw reconstruction, a technique that assigns reconstructed objects to the most probable parts of the decay trees, allowing separation of the signal from the background. This involved theoretical 'scenarios' with large and intermediate mass splitting between the parent particle and lightest supersymmetric particle, as well as the scenario where this mass splitting is low; close to the mass of the Z- boson.

Results were found to be compatible with the standard model expectations for the scenarios with large and intermediate mass splitting. Improved (higher) constraints on supersymmetric partner mass-energies were developed with mass-energies less than 600 GeV excluded at 95% confidence level for the lightest supersymmetric particle.

Possibilities for a potential excess of data above the expected background were found for the scenarios targeting low mass splitting near to the mass–energy of the Z-boson. However, the largest local excess amounts to 3.0 standard deviations; not sufficient to be a signal for new physics. More data are required to determine if this signal is real or just a chance fluctuation.

Ultimately, the aim of physics is to understand the Universe as it exists, not as we would like it to be. Loop quantum gravity, as we have shown in chapter 2, indicates that space–time evolved in the very early universe as a random computational spin network mediated by a weightless spin 2 boson which we identify with the graviton. This occurred within the initial Planck regime and gave rise to a granular form of space–time satisfying Dirac's criterion for finiteness. This implies that the Universe is not fractal and is non-renormalisable. As the Universe continued to expand and cool, additional massive particles were created along with an emergent macroscopic, smooth space–time. We describe how this this might have occurred, by exploiting the mathematics of supersymmetric string theory. This analysis was inspired by reference [12], in particular the contributions of Strominger and Witten in volume 2 on supersymmetry and heterotic strings. Clearly, from the Run 2 results of the LHC, we have to have a very good reason for why supersymmetry has not so far been detected. The evidence from the ATLAS detector shows two things;

(a) Supersymmetry, if it exists, inhabits a high energy realm, like an oasis in the 'desert' between the 13 TeV limit of the LHC and the extreme energy of the Planck regime.
(b) The symmetry is probably broken, due to the high lower limits imposed on the sparticle masses by the ATLAS results.

These results can be explained by assuming a broken form of the symmetry in which the sparticle partner has a much larger mass–energy than its particle 'mate'. As hinted at in the discussion of the ATLAS data, these heavy sparticles could be stable solutions to the puzzle of the missing mass of the Universe called 'dark matter'. Because we have a completely finite, non-perturbative solution for the discrete structure of space–time, this assumption does not cause us any renormalisation problems. As noted in chapter 2, a Wilsonian renormalisation flow has a lower limit causing a perceived 'negative gravitational pressure' at the Planck scale of separation. This may be the driving force behind inflation in the early universe and in theories replacing the 'Big Bang' with a 'cosmic bounce'. These reasons explain much which is currently inexplicable and are consistent with the facts. They thus represent a good guess in terms of our black box model of quantum gravity (see chapter 1).

3.9 Adding fermions and bosons to the mix

The graviton spreads the long-range influence of gravity at the speed of light, thus it must be massless. We need to address the question of how the common massive particles, both fermions and bosons, emerged as the Universe expanded and cooled beyond the Planck regime.

We assume that after the Planck regime, an apparently smooth macroscopic space–time emerged from the initial evolution of a random computational spin network described mathematically in chapter 2. This was followed by the first particles with mass, including the electron. As the young universe continued to expand and cool, the heavier protons and neutrons emerged, forming complex nuclei of deuterium and helium by the end of the first few minutes [13].

We begin our mathematical analysis of this post-Planck regime by recalling the Kaluza–Klein (KK) model. For a metric tensor $g(z)$ defined on an emergent smooth five-dimensional manifold Y we have the Lagrangian action;

$$S(g) = \frac{1}{2\pi k^2}\int_Y \sqrt{-\det(g(z))}\, R(g(z)) d^5 z. \tag{3.4}$$

Here we assume R is the trace of the Ricci tensor; the contracted form of the Riemann curvature tensor; and k is Newton's gravitational constant. The classical vacuum solution is $R(g(z))_{MN} = 0$, with the parameters M, N ranging over 0,1,2,3,4. Assume now that the fifth dimension is compactified, forming a circle of small radius r, parameterized by the variable θ. We choose for Y flat Minkowski space–time 'Euclideanised' by a Wick rotation, crossed with the unit circle so that;

$$Y = \boldsymbol{R}^4 \times S^1.$$

Then we have the five-dimensional covariant metric tensor $g(x, \theta)_{MN} dz^M dz^N = \eta_{\mu\nu} dx^\mu dx^\nu + r^2 d\theta^2$ in an obvious notation, where $\eta_{\mu\nu}$ is the Wick rotated mostly plus form of the four-dimensional covariant metric tensor. Introducing a scalar bosonic field $\rho(x, \theta)$ on the manifold Y we can rewrite this as;

$$g(x, \theta)_{MN} dz^M dz^N = \exp\left(-\frac{\rho}{3}\right)(\exp(\rho)(d\theta + kA_\mu x^\mu)^2 + \eta_{\mu\nu} dx^\mu dx^\nu).$$

Expanding the metric tensor in Fourier coefficients of the circle coefficient yields, with normal ordering;

$$g_{\mu\nu}(x, \theta) = \sum_{n=-\infty}^{\infty} g_{\mu\nu;n}(x) \exp(in\theta).$$

With similar expressions for $A_\mu(x, \theta)$ and $\rho(x, \theta)$. We choose, based on the analysis at chapter 2, the radius of the circle r to be equal to the Planck length $l_P = 10^{-33}$ cm. At this value all of the coefficients n have an energy contribution of at least $\frac{|n|}{r}$. In the low to medium energy regime beyond the standard model we can thus ignore all the Fourier coefficients except for the case $n = 0$. Thus, we have $g_{\mu\nu}(x, \theta) = g_{\mu\nu;0}(x)$ and similarly for $A_\mu(x, \theta)$ and $\rho(x, \theta)$.

We now consider a linear perturbation of the five-dimensional metric tensor of the form;

$$\hat{g}_{MN} = g_{MN} - h_{MN}.$$

The equations of motion from the Lagrangian at equation (3.4) are the Einstein equations $R_{MN}(g) = 0$. Thus we have, in an obvious notation, $R_{MN}(\hat{g} + h) = 0$. Expanding this to first order in h, and combining with the result of the Fourier expansion yields an effective action which is purely four-dimensional. It is expressed in terms of the $n = 0$ Fourier coefficients which we write as;

$$g_0 = g_{\mu\nu;0}(x); \quad A_0(x) \text{ and } \rho_0(x).$$

The resultant effective four-dimensional action is given in [12], for the low to medium energy regime, by;

$$S(g_0(x), A_0(x), \rho_0(x)) = r \int_{\mathbf{R}^4} d^4x \sqrt{-g_0} \left(\frac{1}{k^2} R(g_0) - \frac{1}{4} \exp(\rho_0) F^0_{\mu\nu} F_0^{\mu\nu} - \frac{1}{6k^2} (\nabla \rho_0)^2 \right).$$

In summary, we began with a five-dimensional pure gravity theory by taking the cross-product of four-dimensional space–time with a compactified circle. We discovered that the low-energy effective action can be expressed in four dimensions as the sum of three four-dimensional fields namely; a metric, a connection and a scalar field. The symmetries of this low energy action are those which keep either space–time or the compactified circle group $U(1)$ constant.

3.10 The 10-dimensional pure gravity action

Guided by the simpler five-dimensional case, we now apply the same analysis to a 10-dimensional Einstein–Hilbert pure gravity action of the form

$$S(g) = \int_Y d^{10}x \sqrt{-\det(g)}\, R(g). \tag{3.5}$$

where g is the 'mainly +' 10-dimensional metric tensor on the manifold Y. The solution minimising this action is again the Ricci flat metric satisfying $R_{MN}(g) = 0$ where the Ricci 2-form R_{MN} ($M,N = 0,1,2,\ldots,9$) is the contracted form of the 10-dimensional Riemann curvature tensor.

In analogy with the KK approach we replace the compactified circle of radius equal to the Planck length with a six-dimensional compact set which we denote as K^6. As before we use a Wick rotation to Euclideanise the basic four-dimensional manifold, denoting it as R^4. The manifold Y is then the cross-product $Y = R^4 \times K^6$. We define a local coordinate set ($m,n = 0,1,2,\ldots,5$) on K^6. The metric solution to the minimisation of least action can then be written as a covariant split metric tensor as follows;

$$g_{MN} = g_{\mu\nu} + g_{mn}.$$

If this metric tensor g is Ricci flat on Y it is topologically Ricci flat on the subset K^6. From the discussion in chapter 2, we can represent a gravitational field on Y by adjusting the orientations of the local compact double-cones. Without particles of non-zero mass this is a *pure gravity solution.*

Following the KK analysis, we now consider a small perturbation of the form

$$\hat{g}_{MN} = g_{MN} - h_{MN}.$$

Let $\boxdot^j$ denote the j-dimensional d'Alembertian operator. The full d'Alembertian $\boxdot^{10}$ acting on $Y = R^4 \times K^6$ splits into $\boxdot^4 + \boxdot^6$, with;

$$\boxdot^{10} h_{MN} = \boxdot^4 h_{\mu\nu} + \boxdot^6 h_{mn}$$

where $\boxdot^4 h_{\mu\nu} = \eta^{\alpha\beta}\partial_\alpha\partial_\beta h_{\mu\nu}$ and $\boxdot^6 h_{mn} = (g_0)^{ab}\nabla_a\nabla_b h_{mn}$.

Again following the KK analysis, we expand $h_{\mu\nu}$ in Fourier modes of the harmonics of the compact set K^6 which are scalar functions $f(k)$ acting on K^6 and are eigenstates of the d'Alembertian operator $\boxdot^6$. This means that we have, with normal ordering;

$$h_{\mu\nu} = h_{\mu\nu;k} f_k$$
$$\boxdot^6 f_k = \lambda(k) f_k .$$

3.11 Adding bosons to the theory

We represent bosons by adding to the Einstein–Hilbert action at equation (3.5) a term of the form $S(\varphi)$ where φ is a scalar valued function on 10-dimensional space–time. Expanding $\varphi = \varphi(z)$ in Fourier modes and neglecting all extreme mass modes implies that after compactification φ is a boson field on four-dimensional space–time. Typically, the function S is a polynomial of small degree such as $S(\varphi) = a\varphi^4$ giving the ' φ^4-theory' which has been extensively analysed.

3.12 Adding fermions

This is much trickier. The basic problem is that string theory is essentially a bosonic theory. To make it relevant to fermion states we have to assume that the theory is supersymmetric. Our earlier discussion of the ATLAS data indicates that this symmetry is badly broken, and we will discuss the mathematics of how this might have occurred a little later. For now, we can, as we did for bosons, begin with a 10-dimensional supersymmetric string theory and then collapse down to the four dimensions of normal space–time. We exploit for this the theory of closed orientable superstrings constructed as a chiral combination of the closed 26-dimensional bosonic and 10-dimensional fermionic strings. The resulting *heterotic* 10-dimensional theory is $N = 1$ supersymmetric, and Lorentz invariant. We can proceed using heterotic string theory and, keeping things simple, with K^6 a compact orbifold. The details are clearly laid out in chapter 10 of Bailin and Love [14] and there is no need to repeat them here. Finally, we promised to provide a mathematical discussion of how supersymmetry could have been broken and we do this now. We exploit further the theory of the supersymmetric string and use symmetry breaking to produce mass. The first part of our discussion repeats part of the graviton experience but it is worth checking out the details as we go along.

If we, exploiting relativistic covariance, chose an inertial reference frame in which $P^\mu = (E, 0, 0, E)$ then

$$\{Q_A, \bar{Q}_{\dot{B}}\}|p, \lambda >= \sigma^\mu_{A\dot{B}} P_\mu |p, \lambda >= \sigma^\mu_{A\dot{B}} p_\mu |p, \lambda\rangle.$$

Now

$$\sigma^\mu_{A\dot{B}} p_\mu = \sigma^0 p_0 - \sigma^1 p_1 - \sigma^2 p_2 - \sigma^3 p_3 = E(\sigma^0 - \sigma^3)$$
$$= E\left(\begin{pmatrix}1 & 0\\0 & 1\end{pmatrix} - \begin{pmatrix}1 & 0\\0 & -1\end{pmatrix}\right) = \begin{pmatrix}0 & 0\\0 & 2E\end{pmatrix}.$$

Thus for the $1\dot{1}$ (top left hand) element we have $Q_1\bar{Q}_{\dot{1}} + \bar{Q}_{\dot{1}}Q_1 = 0$.

Applying this to the Clifford vacuum we have;

$$|\langle p, \lambda|Q_1\bar{Q}_{\dot{1}}|p, \lambda\rangle = -\langle p, \lambda|\bar{Q}_{\dot{1}}Q_1|p, \lambda\rangle = 0.$$

Similarly, considering the $2\dot{1}$ element of the matrix, we have

$$|\langle p, \lambda|Q_2\bar{Q}_{\dot{1}}|p, \lambda\rangle = -\langle p, \lambda|\bar{Q}_{\dot{1}}Q_2|p, \lambda\rangle = 0.$$

Thus $(aQ_1 + bQ_2)\bar{Q}_{\dot{1}}|p, \lambda\rangle = 0$ for any scalars a and b. Since Q_1 and Q_2 span the space L_1, it follows that $\bar{Q}_{\dot{1}}|p \cdot \lambda\rangle = 0$.

The only non-trivial state in this irreducible representation, apart from the Clifford vacuum is; thus $\bar{Q}_{\dot{2}}|p, \lambda\rangle$ noting that for the $2\dot{2}$ element of our earlier matrix equation we have; $\{Q_2, \bar{Q}_{\dot{2}}\}|p, \lambda\rangle = \sigma^\mu_{2\dot{2}} p_\mu = 2E \neq 0$.

In normalised form this state is given by $\psi = \frac{1}{2\sqrt{E}}\bar{Q}_{\dot{2}}|p, \lambda\rangle$ and has spin value $\lambda - \frac{1}{2}$. As before this state has the same mass–energy p as the Clifford vacuum state. Thus in this irreducible representation there are just two non-zero states, with one a fermion state and the other a boson state. The effect of the operator $\bar{Q}_{\dot{2}}$ is to lower the spin state by one half unit (acting in this case as an *annihilation operator*), transforming a vacuum boson state into a fermion, or vice versa.

Various values of the Clifford vacuum spin eigenvalue λ produce a range of options.

- $\lambda = \frac{1}{2}$ gives rise to a Weyl spinor fermion as the Clifford vacuum state, with $\psi = \frac{1}{2\sqrt{E}}\bar{Q}_{\dot{2}}|p, \lambda\rangle$ corresponding to a bosonic scalar particle of equal mass. This pairing is known as a *chiral supermultiplet*. As we have demonstrated, the theory predicts that in this chiral representation each particle has a supersymmetric companion particle of equal mass–energy. This case can be applied to quarks, leptons and Higgs particles as follows:

Fermion state $\|p, \lambda\rangle$ with spin $\frac{1}{2}$	Boson state ψ with spin 0
Quark	Squark
Lepton	Slepton
Higgsino	Higgs boson

- Choosing $\lambda = 1$ means that the Clifford vacuum is a Boson state with spin value 1. This applies to gauge bosons (e.g. in the Salaam–Weinberg electroweak theory) with gauginos as their supersymmetric fermion partner states having spin $\frac{1}{2}$.
- Choosing $\lambda = 2$, our previous analysis resulted in a boson state as the Clifford vacuum, and this is a spin-2 graviton (the boson which mediates the gravitational force). Its supersymmetric fermion partner is the Gravitino with spin $\frac{3}{2}$.

Massive (fixed non-zero rest energy) representations of the Z_2-graded Lie algebra can be treated in a similar way, but in general, mass is better treated in the theory as due to a symmetry breaking process as discussed in the next (and concluding) section.

3.13 Symmetry breaking to create mass

We briefly recall here the concept of a superspace from earlier in this chapter. As we have demonstrated above, an irreducible representation of a graded Lie algebra yields an example of a supersymmetric representation corresponding to a twin pair of particles, one a fermion and the other a boson. This graded Lie algebra is defined by algebraic anticommutators of the form $\{Q_A, \bar{Q}_B\} = \sigma^{\mu}_{AB}P_{\mu}$.

By adding a pair of Grassmann variables to this algebraic formulation, we can turn this graded Lie algebra back into a standard Lie algebra, and turn the anticommutators of the graded Lie algebra into normal Lie commutators. The resultant Lie algebra is called a superspace. It has an underlying Lie group G whose generators are the elements of the superspace algebra. A typical unitary element of the group G is thus of the form

$$G(x^{\mu}, \xi, \bar{\xi}) = \exp i(\xi Q + \bar{\xi}\bar{Q} - x^{\mu}P_{\mu}).$$

As a fibre bundle, we can interpret the group product as shifting the locus of the Grassmann vector in space–time together with additive change to the Grassmann vector at this point. If this shift is infinitesimal, then, as in normal Lie group theory, we can consider the tangent plane around the group element $G(x^{\mu}, \xi, \bar{\xi})$. This then gives the following local representation of these Lie group generators:

$$P_{\mu} = i\frac{\partial}{\partial x^{\mu}} = i\partial_{\mu}$$
$$iQ_A = \frac{\partial}{\partial \theta_A} - i\sigma^{\mu}\bar{\theta}\frac{\partial}{\partial x^{\mu}}$$
$$i\bar{Q}_{\dot{A}} = -\frac{\partial}{\partial \bar{\theta}_{\dot{A}}} + i\theta\sigma^{\mu}\frac{\partial}{\partial x^{\mu}}.$$

We constructed the irreducible representation of the graded Lie algebra as quantum operators Q acting on a supersymmetric pair consisting of a single bosonic scalar state and a single fermionic spinor state. To apply a Higgs type symmetry breaking approach in this context we first have to extend this thinking from a single particle state to a quantum field. We consider then a free boson field $\varphi(x)$ corresponding to a scalar field defined at each point x of flat space–time.

We now construct an infinitesimal unitary transformation, based on superspace characteristics, of the form $U(\xi) = \exp(-i\xi.\ Q) \simeq 1 - i\xi \cdot Q$ for an infinitesimal value of the Grassman variables ξ, $\bar{\xi}$. In the superspace Lie algebra, we define the supersymmetric transformation δ_ξ by $\delta_\xi \cdot \phi(x) = [i(\xi \cdot Q), \phi(x)]$ for the free boson field $\varphi(x)$

From the algebraic identity $[\xi Q, \bar{\eta}\bar{Q}] = 2\xi\sigma^\mu\bar{\eta}P_\mu$ it follows that

$$\begin{aligned}[\delta_\eta, \delta_\xi] \cdot \varphi(x) &= (-2(\eta\sigma^\mu\bar{\xi}) + 2(\xi\sigma^\mu\bar{\eta}))[P_\mu, \varphi(x)] \\ &= 2(\xi\sigma^\mu\bar{\eta} - \eta\sigma^\mu\bar{\xi})i\partial_\mu\varphi(x)\end{aligned}$$

taking the tangent plane linear representation for the Lie group generator P_μ and remembering the Euler–Lagrange equation (discussed just below).

Similarly, for a free spinor vector field $\psi_A(x)$, we have;

$$[\delta_\eta, \delta_\xi] \cdot \psi_A(x) = 2(\xi\sigma^\mu\bar{\eta} - \eta\sigma^\mu\bar{\xi})i\partial_\mu\psi_A(x)$$

provided we assume that the Euler–Lagrange dynamical equation $\bar{\sigma}^\mu\partial_\mu\psi_A(x) = 0$ is satisfied.

Solutions of this type, satisfying the Euler–Lagrange equations are called *on-shell solutions.*

The dynamics of these free fields are governed by the Lagrangian $\mathcal{L}$. The Lagrangian of a free massless boson field and a free fermionic Weyl spinor field is given by;

$$(\partial_\mu\phi^*)(\partial_\mu\phi) + i\bar{\psi}_A\sigma^\mu\partial_\mu\psi_A$$

from which we can deduce that the Lagrangian is invariant under the following supersymmetric transformation which essentially swaps the two field variables over;

$$\begin{aligned}\delta_\xi\varphi(x) &= \sqrt{2}\,\xi\psi_\alpha(x) \\ \delta_\xi\psi_\alpha(x) &= i\sqrt{2}\,\sigma^\mu\bar{\xi}\,\partial_\mu\varphi(x).\end{aligned}$$

This is an example of an on-shell solution for a supersymmetric symmetry transformation—i.e. a transformation for which the Lagrangian remains invariant and where the fields satisfy the Euler–Langrage dynamical equations generated by the Lagrangian.

If the Euler–Lagrange equations are not satisfied, we have what is known as an 'off-shell' solution. In this case we can adopt a Higgs-type symmetry breaking approach by adding an auxiliary scalar field $F(x)$ to the Lagrangian to balance the two components of the Weyl spinor field. Then we have

$$\psi_\alpha(x) = \begin{pmatrix}\psi_1(x) \\ \psi_2(x)\end{pmatrix} \text{ is balanced by the scalar fields } \begin{pmatrix}\varphi(x) \\ F(x)\end{pmatrix}.$$

From the basic algebraic relations of the Lie algebra we have;

$$P_\nu = \frac{1}{4}(\bar{\sigma}^\nu)^{\dot{\beta}\alpha}\{Q_\alpha, \bar{Q}_{\dot{\beta}}\}.$$

Thus the Hamiltonian

$$H = P_0 = \frac{1}{4}(\bar{\sigma}^0)^{\dot{\beta}\alpha}\{Q_\alpha, \bar{Q}_{\dot{\beta}}\} = \frac{1}{4}(\{Q_1, \bar{Q}_{\dot{1}}\} + \{Q_2, \bar{Q}_{\dot{2}}\}).$$

Hence H is a hermitian and positive definite quantum operator.

If $|0\rangle$ is a Clifford vacuum state, then, assuming Supersymmetry is not broken,

$$Q_\alpha|0\rangle = 0 \text{ and } \bar{Q}_{\dot{\alpha}}|0\rangle = 0 \text{ for } \alpha = 1, 2 \text{ and } \dot{\alpha} = \dot{1}, \dot{2}.$$

Hence the expected energy of the Clifford vacuum state when supersymmetry is unbroken is given by;

$$\langle 0|H|0\rangle = \langle 0|\frac{1}{4}(\{Q_1, \bar{Q}_{\dot{1}}\} + \{Q_2, \bar{Q}_{\dot{2}}\})|0\rangle = 0.$$

If we add the auxiliary scalar field $F(x)$ to the Lagrangian in such a way that $\langle 0|F(x)|0\rangle$ is strictly greater than 0 then this is sufficient to make the expected energy of the Clifford vacuum state $|0\rangle$ strictly greater than zero; thus supersymmetry is now broken.

The Lagrangian now takes the form;

$$\mathcal{L}(\varphi(x), \psi_A(x)) = (\partial_\mu\varphi(x)^*)(\partial_\mu\varphi(x)) + i\bar{\psi}_A(x)\sigma^\mu\partial_\mu\psi_A(x) + F(x)^*F(x).$$

Note that the auxiliary field F is added in such way that it forms a real scalar field addition to the Lagrangian (as is the Higgs field in the electroweak theory).

This 'F-term symmetry breaking' allows the masses of the fermions and their bosonic partners to differ significantly within the same particle multiplet.

References

[1] Kirrilov A 2008 *An Introduction to Lie Groups and Lie Algebras* (Cambridge: Cambridge University Press)

[2] Connes A 1994 *Noncommutative Geometry* (San Diego, CA: Academic)

[3] Araki H 1999 *Mathematical Theory of Quantum Fields* (Oxford: Oxford University Press)

[4] Moffat J and Wang C 2017 Factorial unitary representations of the translational group, invariant pure states and the supersymmetric graviton *J. Phys. Math.* **8** 257

[5] Moffat J 1975 Connected topological groups acting on von Neumann algebras *J. London Math. Soc.* **2** 411–7

[6] Moffat J 1973 Continuity of automorphic representations *Proc. Camb. Philos. Soc.* **74** 461–5

[7] Hewitt E and Ross K 1963 *Abstract Harmonic Analysis* (Berlin: Springer)

[8] Douglas R G 1972 *Banach Algebra Techniques in Operator Theory* (New York: Academic)

[9] Ehresmann C 1950 *Les Connections Infinitesimales dans un Espace Fibre Differentiable* (Bruxelles: Colloque de Topologie) pp 29–55

[10] Salam A and Strathdee J 1978 Supersymmetry and superfields *Fortschr. Phys.* **26** 57

[11] Aaboud M *et al* (ATLAS Collaboration) 2018 Search for chargino-neutralino production using recursive jigsaw reconstruction in final states with two or three charged leptons in proton-proton collisions at 13 TeV with the ATLAS detector *Phys. Rev.* D **98** 092012

[12] Deligne P *et al* 1999 *Quantum Fields and Strings: A Course for Mathematicians* vols 1, 2 (Providence, RI: American Mathematical Society)

[13] Weinberg S 1988 *The First Three Minutes* (New York: Basic Books) updated version

[14] Bailin D and Love A 1994 *Supersymmetric Gauge Field Theory and String Theory* (London: Taylor and Francis)

IOP Publishing

A Mathematically Coherent Quantum Gravity

James Moffat

Chapter 4

The algebraic nature of reality

Mathematical thought removes the spirit from its worldly haunts to solitude…[1].

4.1 Introduction

From the Preface to this book I extract and repeat the words of A Einstein.

> *One can give good reasons why reality cannot at all be represented by a continuous field. From the quantum phenomena it appears to follow with certainty that a finite system with finite energy can be completely described by a finite set of numbers (quantum numbers). This does not seem to be in accordance with continuum theory and must lead to an attempt to find a purely algebraic theory for the description of reality. But nobody knows how to obtain the basis of such a theory.*

I have constructed the stage in chapter 2 and the players who interact with the stage and each other in chapter 3. In this final theory chapter, before the wrap up in chapter 5, I now seek out and describe some more exotic creatures inhabiting this new algebraic reality, which Einstein could not find.

- I begin with the true nature of Yang–Mills quantum fields as an answer to the Clay Mathematical Institute Millennial question about an underpinning mathematical theory (more detail is given in chapter 5 on the question at issue). This includes the relation between quantum fields and local emergent algebras defined on compact subsets of space-time.
- I then describe the structure of these local algebras based on von Neumann's structure theory for W*-algebras.

[1] From the Preface to the classic work *The Theory of Groups and Quantum Mechanics* by Hermann Weyl.

doi:10.1088/978-0-7503-2580-6ch4

- Finally, I discuss the existence and nature of diffeomorphism invariant states, a concept introduced by Ashtekar in loop quantum gravity.

4.2 Symmetry invariance and symmetry breaking in Yang–Mills quantum fields

A Yang–Mills quantum field theory is defined in terms of a compact non-abelian Lie group G and is thus directly related to the properties of what I call *Wigner sets.* As we will discuss in more detail below, these sets lead to symmetry invariance relative to a non-abelian compact gauge group G (such as $G = SU(n)$). We also discuss how the local algebra of observables $O(D)$ is defined in terms of quantum fields over the subset D of space–time. We can then apply the logic of *F-type* symmetry breaking to the algebra $O(D)$ to show that all non-vacuum quantum states of this algebra are distinct from the Clifford vacuum state by at least a given real value giving rise to a mass gap. I make a start by defining the concept of an emergent local algebra.

4.2.1 Wigner sets and symmetry invariance in the local algebra *O*(*D*)

For each event point $\mathbf{x}$ in Minkowski space–time, we have a fibre algebra $\mathbf{A(x)}$ defined as a von Neumann algebra with trivial centre and a faithful representation as an algebra of operators acting on a separable Hilbert space. During the Planck era we can define a *local algebra* $O(D)$ corresponding to the finite direct sum of these fibre algebras. As the Universe expands and cools beyond the Planck era, this can be approximated by a limiting *direct integral algebra* $O(D)$ *of quantum fields* of a fibre bundle with emergent base space a finite and bounded (compact) Borel subset of space–time, D. The fibre algebras give rise to a section mapping $\mathbf{x} \to \mathbf{A(x)}$ from D to the fibre bundle $O(D)$ which is the emergent von Neumann algebra consisting of linear observables which are now direct integrals of the form $\mathbf{A} = \int_D^{\oplus} \mathbf{A(x)dx}$ as discussed in [1]. The algebraic operations of addition and multiplication are assumed defined *fibrewise* for this algebra. In other words

$$(\mathbf{A} + \mathbf{B} \cdot \mathbf{C})(\mathbf{x}) = \mathbf{A(x)} + B(x) \cdot C(x)$$

in the fibre algebra $\mathbf{A(x)}$ rooted at the event point $\mathbf{x}$.

Thus $O(D)$ is an associative principal fibre subbundle; *associative* in the sense that a Lie group (the compact group G) acts on each fibre $\mathbf{A(x)}$; a *subbundle* in the sense that only that subset of $\{\mathbf{A(x)};\ \mathbf{x} \in M\}$ with $\mathbf{x} \in D$ is of physical interest.

We can analyse the group of automorphisms $\propto(G)$ of the local von Neumann algebra $O(D)$ by considering the action on each fibre algebra $\mathbf{A(x)}$ and then building back up to the local algebra, as described in [1]. $O(D)$ does not necessarily have a trivial centre; its structure is more complex, in general.

Given a normal quantum state f, and a weakly measurable representation $g \to \alpha_g$ of a compact Lie group G as gauge automorphisms of the fibre algebra $\mathbf{A(x)}$, we consider the *induced transformation*;

$$\nu_g(f) = f \circ \alpha_{g^{-1}}.$$

From now on we ignore the inverse used in this definition, to avoid notational clutter. It follows from [2] that the mapping $g \to f \circ \alpha_g$ is both norm and weakly continuous. Define the closed convex hull;

$$X(f) = \bar{co}\{f \circ \alpha_g;\ g \in G\} \subset \mathbf{A(x)}_*$$

the predual of **A(x)**, with closure in the $\sigma(\mathbf{A(x)}_*, \mathbf{A(x)})$-topology. Define the induced group of isometric and $\sigma(\mathbf{A(x)}_*, \mathbf{A(x)})$-continuous transformations mapping $X(f) \to X(f)$ by

$$\nu(G) = \{\nu(g): x \to x \circ \alpha_g;\ g \in G,\ x \in X(f)\}.$$

Mathematically, we note that since G is compact and the mapping $f \circ \alpha: G \to \mathbf{A(x)}_*$ is continuous; this implies that $f \circ \alpha(G)$ is $\sigma(\mathbf{A(x)}_*, \mathbf{A(x)})$-compact. The Krein–Smulian theorem [3] then shows that $X(f)$ is also a $\sigma(\mathbf{A(x)}_*, \mathbf{A(x)})$-compact set. Thus $X(f)$ is a non-void $\sigma(\mathbf{A(x)}_*, \mathbf{A(x)})$-compact convex subset of the locally convex Hausdorff linear topological space of ultraweakly continuous linear functionals acting on the fibre algebra **A(x)**. The group of mappings $\nu(G) = \{\nu(g): x \to x \circ \alpha_g;\ g \in G,\ x \in X(f)\}$ is, as we will now show, a non-contracting (semi)-group of weakly continuous affine maps of $X(f)$ onto itself.

If x, y are in $X(f)$ then;

$$\begin{aligned} \|\nu(g)x - \nu(g)y\| &= \sup_{\|A\| \leqslant 1} \left\| \nu(g)x(A) - \nu(g)y(A) \right\| \\ &= sup_{\|A\| \leqslant 1} \left\| x \circ \alpha_g(A) - y \circ \alpha_g(A) \right\| \\ &= \|x - y\| \end{aligned}$$

since each α_g is a continuous bijection.

Thus if $x \neq y$ and $S =$ the weak closure of $\{\nu(g)x(A) - \nu(g)y(A); A \in \mathbf{A(x)}\}$ then $0 \notin S$. Thus each mapping $\nu(g)$ is non-contracting. Also, each $\nu(g)$ is affine, since for A in **A(x)** and $0 \leqslant \lambda \leqslant 1$, we have;

$$\begin{aligned} \{\nu(g) \circ (\lambda x + (1 - \lambda)y)\}(A) &= \lambda x(\alpha_g(A)) + (1 - \lambda)y(\alpha_g(A)) \\ &= \lambda\nu(g) \circ x(A) + (1 - \lambda)\nu(g) \circ y(A). \end{aligned}$$

We can therefore, apply the Ryll–Nardzewski fixed point theorem [4] to establish the existence of an invariant normal state contained in $X(f)$.

This proof probably also applies to the fibre bundle $O(D)$. The physical implications of this result highlight the role of what I define as *Wigner sets*.

Given $g \in G$, define the *Wigner set* of the map $\nu(g): X(f) \to X(f)$ as the group stabiliser

$$\mathcal{F}(\nu(g)) = \{x \in X(f); \nu(g) \circ x = x\}.$$

More generally, given a finite subset $\{g(j) \in G; j = 1, 2, \ldots n\}$ and corresponding mappings $\{\nu(g(j)); j = 1, 2, \ldots n\}$, construct the affine mapping $\frac{1}{n}\left(\sum_{j=1,\ldots,n} \nu(g(j))\right): X(f) \to X(f)$.

We then define

$$\mathcal{F}\{\nu(g(j)); j = ,2, \ldots n\} \triangleq \mathcal{F}\left(\frac{1}{n}\left(\sum_{j=1,\ldots,n} \nu(g(j))\right)\right). \tag{4.1}$$

We assert that, given definition (4.1), the following relationship between Wigner sets applies;

$$\bigcap_j \mathcal{F}(\nu(g(j))) = \mathcal{F}\left(\frac{1}{n}\left(\sum_j \nu(g(j))\right)\right). \tag{4.2}$$

Clearly if $x \in \bigcap_{j=1,\ldots,n} \mathcal{F}(\nu(g(j))) \Rightarrow \nu(g(j)) \circ x = x \; \forall\, j = 1, \ldots, n \Rightarrow$

$$\frac{1}{n}\left(\sum_{j=,\ldots,n} \nu(g(j))\right) \circ x = x$$

$$\Rightarrow x \in \mathcal{F}\left(\frac{1}{n}\left(\sum_j \nu(g(j))\right)\right) \text{ therefore } \bigcap_j \mathcal{F}(\nu(g(j))) \subseteq \mathcal{F}\left(\frac{1}{n}\left(\sum_j \nu(g(j))\right)\right).$$

To complete the proof of equation (4.2) we now need to show that;

$$\mathcal{F}\left(\frac{1}{n}\left(\sum_j \nu(g(j))\right)\right) \subseteq \bigcap_j \mathcal{F}(\nu(g(j))). \tag{4.3}$$

A proof of this is given as part of appendix A of my PhD thesis [5], based on reference [4].

4.2.2 Symmetric Yang–Mills quantum states

It is now easy to prove that $X(f)$ contains a symmetric normal Yang–Mills quantum state for the group of isometric and $\sigma(A(x)_*, A(x))$-continuous transformations

$$\nu(G) = \{\nu(g)(x) = x \circ \alpha_g;\; g \in G,\, x \in X(f)\}.$$

Since $\left(\frac{1}{n}\left(\sum_j \nu(g(j))\right)\right)$ is a $\sigma(A(x)_*, A(x))$-continuous affine mapping on the compact convex set $X(f)$ it has a fixed point x (applying Schauder's fixed point theorem). Then $x \in \mathcal{F}\left(\frac{1}{n}\left(\sum_j \nu(g(j))\right)\right)$.

The expression;

$$\bigcap_{j=1,\ldots,n} \mathcal{F}(\nu(g(j))) = \mathcal{F}\left(\frac{1}{n}\left(\sum_{j=1,\ldots,n} \nu(g(j))\right)\right).$$

Shows that the Wigner sets $\mathcal{F}(\nu(g))$ have the finite intersection property since $\bigcap_j \mathcal{F}(\nu(g(j)))$ is non-void. Clearly each Wigner set is a $(\sigma(\mathbf{A}(\mathbf{x})_*, A(x)))$ closed subset of the compact set $X(f)$ thus $\cap_g \mathcal{F}(\nu(g)) \neq \emptyset$.

If $h \in \cap_g \mathcal{F}(\nu(g)) \Rightarrow h = \nu_g \circ h = h \circ \alpha_g \ \forall \ g \in G$. Thus, h is the required invariant quantum state.

The proof shows that quantum states invariant under the action of compact Lie groups are *common* in the sense that the weakly closed convex hull of every normal quantum state contains such an invariant state. If our initial quantum state is faithful then the final quantum state is also faithful.

4.2.3 Superspace

From chapter 3 we have that an irreducible representation of a graded Lie algebra yields an example of a supersymmetric representation corresponding to a twin pair of particles, one a fermion and the other a boson. This graded Lie algebra is defined by algebraic anticommutators of the form $\{Q_A, \bar{Q}_{\dot{B}}\} = \sigma^{\mu}_{A\dot{B}} P_\mu$. Following the ideas of Salam [6], by adding a pair of Grassmann variables to this algebraic formulation, we can turn the graded Lie algebra back into a standard Lie algebra, and turn the anticommutators of the graded Lie algebra into normal Lie commutators. The resultant Lie algebra is called a superspace. It has an underlying Lie group G' whose generators are the elements of the superspace algebra. A typical unitary element of the group G' is thus of the form;

$$G'(x^\mu, \xi, \bar{\xi}) = \exp i(\xi Q + \bar{\xi}\bar{Q} - x^\mu P_\mu).$$

As a fibre bundle, we can interpret the group product as a shifting of the locus of the Grassmann vector in space–time together with additive change to the Grassmann vector at this point. If this shift is infinitesimal, then, as in normal Lie group theory, we can consider the tangent plane at the group element $G'(x^\mu, \xi, \bar{\xi})$. This then gives the following local representation of these Lie group generators:

$$\begin{aligned} P_\mu &= i\frac{\partial}{\partial x^\mu} \\ iQ_A &= \frac{\partial}{\partial \xi_A} - i\sigma^\mu \bar{\xi}\frac{\partial}{\partial x^\mu} \\ i\bar{Q}_{\dot{A}} &= -\frac{\partial}{\partial \xi_A} + i\xi\sigma^\mu \frac{\partial}{\partial x^\mu}. \end{aligned}$$

In this way we can construct an irreducible representation of the $N = 1$ graded Lie algebra as quantum operators Q_A. To apply a symmetry breaking approach in this context we first have to extend this thinking from a single particle state to a quantum field.

We consider then a free boson field $\varphi(x)$ corresponding to a scalar field defined at each point x of flat space–time. We now construct an infinitesimal unitary transformation, based on superspace characteristics, of the form $U(\xi) = \exp(-i\xi \cdot Q) \simeq 1 - i\xi \cdot Q$ for an infinitesimal value of the Grassmann variables $\xi, \bar{\xi}$.

In the superspace Lie algebra, we define the supersymmetric transformation δ_ξ by $\delta_\xi \cdot \phi(x) = [i(\xi \cdot Q), \phi(x)]$ for the free boson field $\varphi(x)$.

From the algebraic identity $[\xi Q, \bar{\eta}\bar{Q}] = 2\xi\sigma^\mu\bar{\eta}P_\mu$ it follows that

$$[\delta_\eta, \delta_\xi].\ \varphi(x) = (-2(\eta\sigma^\mu\bar{\xi}) + 2(\xi\sigma^\mu\bar{\eta}))[P_\mu, \varphi(x)] = 2(\xi\sigma^\mu\bar{\eta} - \eta\sigma^\mu\bar{\xi})i\partial_\mu\varphi(x)$$

taking the tangent plane linear representation for the Lie group generator P_μ and assuming that the Euler–Lagrange equation is satisfied.

Similarly, for a free spinor vector field $\psi_A(x)$, we have;

$$[\delta_\eta, \delta_\xi].\ \psi_A(x) = 2(\xi\sigma^\mu\bar{\eta} - \eta\sigma^\mu\bar{\xi})i\partial_\mu\psi_A(x).$$

Solutions of this type, satisfying the Euler–Lagrange equation are called *on-shell solutions.*

The dynamics of these free fields are governed by the Lagrangian L. Applied to the Lagrangian of a free massless boson field and a free fermionic Weyl spinor field, this is given by;

$$(\partial_\mu\phi^*)(\partial_\mu\phi) + i\bar{\psi}_A\sigma^\mu\partial_\mu\psi_A$$

from which we can deduce that the Lagrangian is invariant under the following supersymmetric transformation which essentially swaps the two field variables over;

$$\delta_\xi\varphi(x) = \sqrt{2}\,\xi\psi_\alpha(x)$$
$$\delta_\xi\psi_\alpha(x) = i\sqrt{2}\,\sigma^\mu\bar{\xi}\,\partial_\mu\varphi(x).$$

This is an example of an on-shell solution for a supersymmetric transformation—i.e. a transformation for which the Lagrangian remains invariant and where the fields satisfy the Euler–Langrage dynamical equations generated by the Lagrangian.

If the Euler–Lagrange equation is not satisfied, we have what is known as an 'off-shell' solution. In this case we can adopt a symmetry breaking approach by adding an auxiliary scalar field $F(x)$ to the Lagrangian to balance the two components of the Weyl spinor field. Then we have

$$\psi_\alpha(x) = \begin{pmatrix} \psi_1(x) \\ \psi_2(x) \end{pmatrix} \text{ is balanced by the scalar fields } \begin{pmatrix} \varphi(x) \\ F(x) \end{pmatrix}.$$

From the basic algebraic relations of the Lie algebra we have;

$$P_\nu = \frac{1}{4}(\bar{\sigma}^\nu)^{\dot{\beta}\alpha}\{Q_\alpha, \bar{Q}_{\dot{\beta}}\}.$$

Thus the Hamiltonian

$$H = P_0 = \frac{1}{4}(\bar{\sigma}^0)^{\dot{\beta}\alpha}\{Q_\alpha, \bar{Q}_{\dot{\beta}}\} = \frac{1}{4}(\{Q_1, \bar{Q}_{\dot{1}}\} + \{Q_2, \bar{Q}_{\dot{2}}\}).$$

Hence H is a hermitian and positive definite quantum operator—i.e. H is an observable.

If $|0\rangle$ is a Clifford vacuum state, then, assuming supersymmetry is not broken,

$$Q_\alpha|0\rangle = 0 \text{ and } \bar{Q}_{\dot{\alpha}}|0\rangle = 0 \text{ for } \alpha = 1, 2 \text{ and } \dot{\alpha} = \dot{1}, \dot{2}.$$

Hence the expected energy of the Clifford vacuum state when supersymmetry is unbroken is given by;

$$\langle 0 \mid H \mid 0 \rangle = \langle 0 \mid \frac{1}{4}(\{Q_1, \bar{Q}_{\dot{1}}\} + \{Q_2, \bar{Q}_{\dot{2}}\}) \mid 0\rangle = 0.$$

If we add the auxiliary scalar field $F(x)$ to the Lagrangian in such a way that

$$\langle 0|F(x)|0\rangle \text{ is strictly greater than } 0$$

then this is sufficient to make the expected energy of the Clifford vacuum state $|0>$ strictly greater than zero; thus supersymmetry is broken. For the local algebra $O(D)$. The Lagrangian now takes the form below since the product or sum of two elements of the dual space of $O(D)$ are defined pointwise;

$$L(\varphi(x), \psi_A(x)) = (\partial_\mu\varphi(x)^*)(\partial_\mu\varphi(x)) + i\bar{\psi}_A(x)\sigma^\mu\partial_\mu\psi_A(x) + F(x)^*F(x).$$

This corresponds to a section through a vector bundle in the dual space $O(D)^*$ with the *action* creating a path in space–time mapping to a section in the vector bundle.

Note that the auxiliary field F is added in such way that it forms a real scalar field addition to the Lagrangian (as is the Higgs field in the electroweak theory). This 'F-term symmetry breaking' allows the masses of the fermions and their bosonic partners to differ significantly within the same particle multiplet, consistent with the experimental results from the ATLAS detector discussed in chapter 3 for run 2 of the LHC running at 13 TeV.

Summarising our discussion thus far

For each event point $\mathbf{x}$ in Minkowski space–time, we have a fibre algebra $\mathbf{A(x)}$ defined as a von Neumann algebra with trivial centre and a faithful representation as an algebra of operators acting on a separable Hilbert space. Earlier in this chapter we defined a *local algebra O(D)* corresponding to the limiting *direct integral algebra of quantum fields* of a fibre bundle with emergent base space a finite and bounded (compact) Borel subset of space–time, D. The fibres give rise to a section $\mathbf{x} \rightarrow \mathbf{A(x)}$ from D to the fibre bundle $O(D)$ as the von Neumann algebra consisting of linear observables which are direct integrals of the form $\mathbf{A} = \int_D^{\oplus} \mathbf{A(x)}d\mathbf{x}$. The algebraic operations of addition and multiplication are assumed defined *fibrewise* for this algebra. Thus $O(D)$ is an associative principal fibre subbundle; *associative* in the sense that a compact Lie group (the group G) acts on each fibre $\mathbf{A(x)}$; a *subbundle* in the sense that only that subset of $\{\mathbf{A(x)}; \mathbf{x} \in M\}$ with $\mathbf{x} \in D$ is of physical interest.

The local von Neumann algebra $O(D)$ can thus be considered as an *algebra of quantum fields*. For the case of a Lagrangian consisting of a spinor field and a scalar field, we constructed an example of a supersymmetric local algebra with F-type symmetry breaking. This corresponds to a lifting from a path in space–time to such a vector bundle in the dual space of the algebraic fibre bundle. We propose to use the ideas of non-commutative ergodic theory to gain further insight into the structure of $O(D)$, as we now describe.

4.3 The structure of local algebras of Yang–Mills quantum fields

J von Neumann introduced the idea of equivalence of measurement projection operators as a way of gaining traction on the structure of a general von Neumann algebra, see for example [7]. Much of this analysis focusses on the question of whether or not the algebra possesses a finite trace, extending the idea of the trace of a finite matrix operator as the sum of its measurement eigenvalues. The analysis was enhanced by E Stormer to take account of groups of (unitarily implemented) automorphisms of the algebra in reference [8]. This allowed him to define a *G-equivalence* of projections which generalises to the non-commutative quantum case the definition Hopf used in standard commutative ergodic theory [9].

I independently developed similar ideas [5], leading to a 1974 paper [10] in which I extended this work to a characterisation of the tensor product of *G-type III* 'entangled' algebras for a general group G. As a result of this previous work we can develop a classification of the structure of our local algebra $O(D)$. We do this by applying these earlier results where the group concerned is now the compact Lie group G. We recall the following definition. Let α: $g \to \alpha_g$ be a representation of the compact group G as a group of automorphisms acting on the von Neumann algebra $O(D)$. A representation π of $O(D)$ acting on a Hilbert space H is *covariant* if there is a group homomorphism $g \to U_g$ from G to the group of unitary operators on H with

$$\pi(\alpha_g(A)) = U_g\pi(A)U_g^* \quad \forall\, A \in O(D).$$

If φ is a normal state of $O(D)$ then $\varphi \circ \alpha_g$ is also a normal state since each automorphism preserves the algebraic structure and hence preserves complete additivity. If S denotes the set of all normal quantum states of $O(D)$ then the direct sum $\pi = \oplus\{\pi_\varphi;\, \varphi \in S\}$ of their Gelfand–Naimark–Segal (GNS) representations is a faithful representation of $O(D)$ as a von Neumann algebra acting on a Hilbert space H which is the direct sum of the GNS Hilbert spaces. If we define $U_g(\oplus_{\varphi\in S}\pi_\varphi(A_\varphi)x_\varphi) = \oplus_{\varphi\in S}\pi_\varphi(\alpha_g(A_{\varphi\,\circ\,\alpha_g})x_\varphi)$ within $\pi = \oplus\{\pi_\varphi;\, \varphi \in S\}$ the GNS construction, then U_g extends to a unitary operator on H and the representation is a faithful normal representation of $O(D)$. We can therefore assume that G acting on $O(D)$ as a group of automorphisms is covariant. From reference [2], since G is a (locally) compact group we can assume continuity in an appropriate topology by making minimal assumptions on measurability of the representation $\alpha(G)$.

We now define the following. Let α: $g \to \alpha_g$ be a group representation of the compact Lie group G as a group acting on the von Neumann algebra $O(D)$. If E and F are projections in $O(D)$ we say that E and F are *G-equivalent* if there is a set of operators $\{A_g; g \in G,\, A_g \in O(D)\}$ with $E = \sum_g A_g^* A_g$ and $\mathrm{F} = \sum_g \alpha_g(A_g A_g^*)$. As noted earlier, this is a direct non-commutative extension of the definition by Hopf [9], in classical ergodic theory. As part of the proof of the main theorem in [10], I developed a representation of the crossed product algebra $\mathscr{R} \times G$ using an entangled tensor product approach. Although not mentioned in [10], I subsequently hypothesised that:

The structure of $\mathscr{R}$ is G-type III if and only if the structure of $\mathscr{R} \times G$ is type III.

We can prove that this hypothesis is true for G a compact Lie group as shown in corollary 4.12 of [5]. This provides a deep insight into the structure of the local algebra $O(D)$ of Yang–Mills quantum fields.

4.4 Developing a diffeomorphism invariant theory for quantum states

Diffeomorphism invariance of quantum states is a key symmetry in quantum gravity and new results were presented by me recently [11]. Within the resulting fibre bundle construct defined in reference [11], our focus is on the local fibre algebra **A(x)** which we assume, as before, to be a factor, isomorphic to a von Neumann algebra with trivial centre, acting on a separable Hilbert space. We now consider the relativistic abelian group T consisting of translations of space–time as a gauge group of automorphisms of **A(x)**.

Let α: $g \rightarrow \alpha_g$ be a continuous representation of T as automorphisms of **A(x)** and f be a faithful normal state of **A(x)**. Also as previously, we consider the *induced transformation;*

$$\nu_g(f) = f \circ \alpha_{g^{-1}},$$

and we ignore the inverse used in this definition.

This construction implies that $\nu_g(f)$ is also a normal state and $\nu_g(f)(A) = f(\alpha_g(A))$. Since the subgroup T is abelian, and the mapping $g \rightarrow \nu(g)$ is a group homomorphism, the set $\{\nu(g); g \in T\}$ is a continuous group of commuting transformations of the dual space **A(x)***.

If f is a quantum state of this fibre algebra of observables as a C*-algebra, then define $\mathcal{E}$ to be the weak* closed convex hull of the set $\{\nu(g)f; g \in T\}$. Because the unit ball of the dual space is weak*-compact, $\mathcal{E}$ is a weak* compact convex set and each $\nu(g)$: $\mathcal{E} \rightarrow \mathcal{E}$. By the Markov–Kakutani fixed point theorem [12], it follows that $\mathcal{E}$ has an invariant element. In other words, the group T has the fixed point property [13], and is thus *amenable.*

Summary

In summary, because T is an abelian group and locally compact it is an amenable group, since the weak*-closed convex hull of any quantum state of the system contains a T-invariant state. This leads us to define the following.

Let **A(x)** be a fibre algebra and consider the group T of translations of space–time. Let α: $g \rightarrow \alpha_g$ be a representation of T as automorphisms of **A(x)**. The group representation α *acts ergodically* on **A(x)** if given a projection E in **A(x)**, $\alpha_g(E) = E \;\; \forall\, g \in T$ implies that $E = 0$ or $E = I$.

This definition is a direct generalisation of the classical, commutative case where **A(x)** is the commuting set of essentially bounded measurable functions on a locally compact space with a regular Borel measure. We also have the following noncommutative generalisation of an ergodic probability measure as an extreme point of the set of invariant measures, as first pointed out by Segal in [14].

Let **A(x)** be a fibre algebra and α: $g \to \alpha_g$ a representation of the translation subgroup T of the Poincaré group P as automorphisms of **A(x)**. A quantum state f of A(x) is *α-invariant* if $f(\alpha_g(A)) = f(A) \quad \forall\ A$ in **A(x)**. If f is a normal (density matrix) state and an extreme point of the set of invariant states, then f is defined to be a *T-ergodic quantum state.*

Note that the set of invariant states is a compact convex subset of the generalised quantum state space of **A(x)** and is thus generated by its extreme points [7]. There is a non-trivial invariant state for the amenable group T thus there is an extremal T-invariant state of **A(x)**. Since, by definition, the Hilbert space representation on which **A(x)** acts is separable, the algebra contains a faithful normal quantum state, and hence a faithful T-invariant normal quantum state. The norm limit of a set of normal states is again normal [7] and thus for a separable Hilbert space, *the fibre algebra* **A(x)** *with the assumptions above always contains a faithful T-ergodic quantum state.*

The algebra **A(x)** is a quantum operator algebra and thus has an identity operator I. If f is a quantum state of **A(x)** then by definition, $f(I) = 1$. *The support of the function* f is the unique smallest projection E in **A(x)** such that $f(E) = 1$ and is denoted E_f. The quantum state f is *faithful* if and only if the support of f equals I.

Now let **A(x)** be a fibre algebra and α: $g \to \alpha_g$ a representation of the translation subgroup T of the Poincaré group P as (gauge) automorphisms of **A(x)**. Let the quantum state f of **A(x)** be faithful, normal and α-invariant. Then f is a faithful normal, ergodic state if and only if the representation α acts ergodically.

Proof. Let π be the Gelfand–Naimark–Segal (GNS) representation of **A(x)** induced by the state f on the Hilbert space $H(f)$. We have that f is a faithful state, thus the corresponding representation π is an isomorphism and the von Neumann algebra π (**A(x)**) has a separating-generating vector ξ.

Define the unitary group $U_g\pi(A)\xi = \pi(\alpha_g(A))\xi$ on a dense subset of $H(f)$, then each U_g extends to a unitary on $H(f) = \{\pi(B)\xi; B \text{ in the fibre algebra}\}^-$ where $\{\}^-$ denotes closure of the set in the appropriate topology. The mapping $U{:}g \to U_g$ is then a unitary representation of the translation group with $U_g\pi(B)U_g^* = \pi(\alpha_g(B))$ $\forall\ g \in T$ and for all B quantum observables in the fibre algebra **A(x).** In other words, the unitary representation U *implements* the automorphic representation $\alpha{:}g \to \alpha_g$.

Consider now the involution mapping on π(**A(x)**) defined as $A \to A^*$. This induces an anti-linear mapping on a dense subset of the Hilbert space; $S{:}A\xi \to A^*\xi$. Moreover, this extends to a mapping with closed graph which we also denote by S. By the theorem of Tomita–Takesaki [7]; S has a polar decomposition $S = J\Delta^{\frac{1}{2}}$ such that, up to isomorphism,

$J(A(x))J = \pi(A(x))'$, the commutant of the fibre algebra $\pi(A(x))$.

If $x = B\xi$ is in the domain of S, then it follows that $U_g x$ also lies in the domain of S, and we have the relationship;

$$U_g SB\xi = U_g B^*\xi = \alpha_g(B^*)\xi = \alpha_g(B)^*\xi = S\alpha_g(B)\xi = SU_g B\xi.$$

This leads to the conclusion that, on the domain of S, we have $U_g S = S U_g$.

Then we have;

$$S = U_g S U_g^* = U_g J \Delta^{\frac{1}{2}} U_g^* = U_g J U_g^* U_g \Delta^{\frac{1}{2}} U_g^*.$$

By uniqueness of the polar decomposition, $J = U_g J U_g^*$; J and U_g commute for all $g \in T$. From this we deduce that;

$$B \in \pi(A(x)) \bigcap \{U_g;\, g \in T\}'$$

implies that $JBJU_g = U_g JBJ$ for all $g \in T$.

Thus $JBJ \in \{U_g;\, g \in T\}' \cap \pi(A(x))'$.

Conversely, if $C \in \{U_g;\, g \in T\}' \cap \pi(A(x))'$ then $C = JBJ$ for some $B \in \pi\,(\mathbf{A(x)})$ and $JBJU_g = U_g JBJ$ implies $JBU_g J = JU_g BJ$ and thus $BU_g = U_g B$ so that $B \in \{U_g;\, g \in T\}'$.

We conclude that

$$J\{\pi(\mathbf{A(x)}) \bigcap \{U_g;\, g \in T\}'\}J = \pi(\mathbf{A(x)})' \bigcap \{U_g;\, g \in T\}'.$$

The automorphic representation α: $g \to \alpha_g$ of T *acts ergodically* if and only if $\pi(\mathbf{A(x)}) \cap \{U_g;\, g \in T\}'$ is trivial, containing only the projections 0 and I? and thus consisting of the set of complex multiples of I. From the reasoning above, and noting that $J^2 = 1$, it follows that the representation α: $g \to \alpha_g$ of T acts ergodically if and only if $\pi(\mathbf{A(x)})' \cap \{U_g;\, g \in T\}'$ is also trivial.

If E is a projection in the set $\pi(\mathbf{A}(x))' \cap \{U_g;\, g \in T\}'$ we can define a quantum state $f_E(A) = \frac{\langle \xi, E\pi(A)\xi \rangle}{\langle \xi, E\xi \rangle}$ on the fibre algebra **A(x).** Then $f_E = \omega_{E\xi} \circ \pi$ is a state dominated by $f = \omega_\xi \circ \pi$ and we have

$$f(A) = \frac{\langle \xi, \pi(A)\xi \rangle}{\langle \xi, \xi \rangle} = \lambda \frac{\langle \xi, E\pi(A)\xi \rangle}{\langle \xi, E\xi \rangle} + (1 - \lambda)\frac{\langle \xi, (I - E)\pi(A)\xi \rangle}{\langle \xi, (I - E)\xi \rangle} \text{ for } A \in \mathrm{A}(x),$$

where

$$\lambda = \frac{\langle \xi, E\xi \rangle}{\langle \xi, \xi \rangle} = \frac{\|E\xi\|^2}{\|\xi\|^2} \text{ and } 1 - \lambda = \frac{\|\xi\|^2 - \|E\xi\|^2}{\|\xi\|^2} = \frac{\langle \xi, (I - E)\xi \rangle}{\|\xi\|^2}.$$

Thus f is an extremal invariant state if and only if the projection $E = 0$ *or* I. The result follows.

In the next section we develop and prove a non-commutative, quantum, version of a result in classical ergodic theory and use it to characterise the existence of such a faithful normal invariant quantum state.

4.4.1 Wandering projections and diffeomorphism invariant quantum states

Hajian and Kakutani, in reference [15], defined *a wandering set* in a measure space as follows;

Let (X, B, μ) be a measure space with finite measure; $\mu(X) < \infty$ and where B is the set of all measurable subsets of X. Let T be a bijective transformation of X such that both T and its inverse are measurable mappings. A *wandering set for* T is a measureable subset S of X such that the sets $\{T^{nk}(S)\}$ are disjoint, for some infinite sequence of integers nk.

Two measures ν and μ on the measure space X are said to be *equivalent* if they share the same null sets. A measure ν is *T-invariant* if $\nu(T(E)) = \nu(E)$ for all measurable subsets E of X.

With these definitions, reference [15] showed that there is a finite T-invariant measure ν on X, equivalent to μ, if and only if there are no wandering subsets of X.

If now we consider an abelian von Neumann algebra $\mathcal{R}$, then $\mathcal{R}$ is isomorphic to $C(X)$ with X a compact Stonean space of finite measure, and the positive, normal, regular Borel measures on X correspond to the normal states of $\mathcal{R}$. By [16] we can characterise these normal measures as being equivalent to measures which annihilate each nowhere dense subset of X. It follows that if measures ν and μ on the measure space X are equivalent and measure ν is normal, then measure μ is also normal.

If θ is a continuous automorphism of the abelian algebra $\mathcal{R}$, isomorphic to $C(X)$, then we can define the homeomorphism T of X by $\theta f(x) = f(Tx)$ for $f \in C(X)$. By the result quoted above, if μ is a normal measure on X with support equal to X, there is a measure equivalent to μ which is T-invariant if and only if there are no wandering measurable subsets E of X.

If such a set E did exist, such that the sets $\{T^{nk}(E)\}$ are disjoint, for some finite or infinite sequence of integers nk, then by regularity of μ we can assume that E is closed. Since X is a Stonean space, we can assume in fact that E is both open and closed. Thus the characteristic function χ_E corresponds to a projection in the algebra $\mathcal{R}$ and the set of projections $\theta^{nk}(\chi_E)$ is an orthogonal set. From the algebraic perspective then we can say the following. Let us assume we are given an abelian von Neumann algebra $\mathcal{R}$, an automorphism θ of $\mathcal{R}$ and a faithful normal state acting on $\mathcal{R}$. Then there is a faithful normal θ-invariant state acting on $\mathcal{R}$ if and only if there are no non-trivial projections E in $\mathcal{R}$ such that for some infinite sequence of integers nk, the projections $\theta^{nk}(E)$ are mutually orthogonal. It can be easily shown that for a commutative algebra this condition on the set of projections is equivalent to the requirement that that there are no nonzero projections E with $\theta^{nk}(E) \to 0$ in the ultraweak topology as $nk \to \infty$ for some infinite sequence nk of integers. This new formulation now generalises easily to the non-commutative, quantum, case as follows.

Let $\mathcal{R}$ be a von Neumann quantum algebra, G a group of automorphisms of $\mathcal{R}$. Then a nontrivial projection E in $\mathcal{R}$ is *wandering* if E is such that; $g_{nk}(E) \to 0$ for some infinite sequence g_{nk} in G. Convergence is defined in the ultraweak operator topology.

If **A(x)** is a fibre algebra then it is a von Neumann quantum algebra with trivial centre and is countably decomposable. Let α: $g \to \alpha_g$ be a group representation of the translation subgroup T of the Poincaré group as automorphisms of **A(x)** which is ultraweakly continuous.

Note concerning normal quantum states

Before continuing, the attendant reader will have remarked on my exploitation of *normal* quantum states, particularly in this chapter. Such states, of course, have a topology which is intertwined with that of the algebra of observables which form a von Neumann or W*-algebra. More than this, however, is the fact that such states follow the rules of quantum logic, as expressed through the lattice of projections of the W*-algebra.

We will now show that there is a faithful normal *diffeomorphism invariant quantum state* on the fibre algebra **A(x)** if and only if there are no wandering projections in **A(x)**.

Proof. Clearly if E is a projection in **A(x)** such that $\alpha_{g_{nk}}(E) \to 0$ for some infinite sequence g_{nk} in T and f is a faithful, normal α-invariant state, then $f(E) = 0$, thus $E = 0$.

The proof of the converse is based on work by M Takesaki on singular states [17, 18]. This emerged from a discussion which I had with Takesaki at the joint CERN-quantum theory experts 'Enrico Fermi' conference at the Villa Monastero in 1973, if I recall it correctly.

We assume that there are no wandering projections in **A(x)**. We have to show that the fibre algebra **A(x)** has a faithful normal T invariant state. We know that the algebra **A(x)** acts on a separable Hilbert space, is countably decomposable and thus has a faithful normal state which we denote as f. Since the group T is amenable, the fixed point property, applied to the set $\mathcal{E}$ defined as the weak* closed convex hull of $\{\nu(g)f;\ g \in T\}$, implies that **A(x)** has an invariant state which we denote as h, which is an element of $\mathcal{E}$. We will show that h is both normal and faithful. By the results of [17, 18], h has a unique decomposition $h = h_n + h_s$ with h_n a normal positive linear functional and h_s a singular positive linear functional. By uniqueness of the decomposition, both of these linear functionals are also α-invariant. Let E be the support of h_n. If $E = I$ then we are done.

Suppose that $E \neq I$ then we will obtain a contradiction. We can choose a projection F with $0 < F < I - E$ and $h_s(F) = 0$. This then implies that we have $h(F) = 0$. Now h is an element of the set $\mathcal{E}$. Thus, from the construction of $\mathcal{E}$ as a weak*-closed convex set there is a sequence φ_n of elements of $\mathcal{E}$ with;

$$\lim_{n\to\infty} \varphi_n(F) = h(F) = 0.$$

It follows that there is a sub-sequence g_{nk} with $f \circ \alpha_{g_{nk}}(F) \to 0$. Since f is both a faithful and normal linear functional this implies that $\alpha_{g_{nk}}(F) \to 0$ in the ultraweak operator topology; i.e. F is a wandering projection. This contradiction shows that the support of h_n equals I and h_n is the required faithful, normal diffeomorphism invariant quantum state.

The same proof also probably works at the level of local algebras. Alternatively, we can build from the fibre algebras to the local algebra as described in [1].

4.5 The information dynamics of black holes

In classical thermodynamics and ergodic theory, entropy is defined through considering the phase space of a dynamical system. The emergent behaviour of such a classical system gives rise to regions of phase space, each corresponding to similar macro-level behaviour. The number and variation in size of these regions reflects the overall complexity of the system. This identification is known as 'coarse graining' of the phase space. The entropy of such a coarse-grained region is essentially a count of all of the different micro-configurations constituting that region; i.e. it is a measure of the volume V of that region in phrase space. Boltzmann's formula for entropy is given by $k \log V$ where k is Boltzmann's constant with the logarithm ensuring that entropy is additive across phase space configurations. The structure of the phase space is such that each set of initial conditions (x,p) generates a unique solution $S(x,p)$.

A solution starting in a low entropy state will tend to wander into larger coarse-grained volumes. Hence thermodynamic entropy tends to increase over time if the system is isolated; giving rise to the second law of thermodynamics. We can see that the law is more valid in a statistical sense when applied to an ensemble of identical systems; the expected entropy of the ensemble will drift to higher values over time.

J von Neumann defined a quantum version of entropy as follows: Let f be a faithful normal quantum state of a local algebra of observables $O(D)$ acting on a separable Hilbert space H. Then for any operator A in $O(D)$ we have $f(A) = \sum_j \langle x_j, Ax_j \rangle$ where the vectors $\{x_j; j = 1, 2, 3, \ldots\}$ form an orthogonal set with $\sum_j \|x_j\|^2 = 1$.

We can write this as $f = \sum_j p_j^2 \omega_{y_j}$ where ω_{y_j} is a vector state with $\|y_j\| = 1$, and $\sum_j p_j^2 = \sum \|x_j\|^2 = 1$.

Von Neumann's non-commutative equivalent of a partition used in classical ergodic theory [5] is the density operator $\rho = \sum_k p_k^2 y_k >< y_k$. This is the weighted sum of the minimal projections E_k onto the vector spaces $E_k H = [|y_k>]$ Then we have the well-known equivalence;

$$
\begin{aligned}
f(A) &= \sum_j p_j^2 \langle y_j, Ay_j \rangle = \sum_{j,m,n} p_j^2 \langle y_j, m\rangle\langle m|A|n\rangle\langle n, y_j\rangle \text{with} \{m, n\} \text{ a discrete basis;} \\
&= \sum_{j,m,n} \langle n, p_j^2 y_j\rangle\langle y_j, m\rangle\langle m|A|n\rangle = \sum_{m,n} \langle n|\rho|m\rangle\langle m|A|n\rangle = \sum_{j,n} \langle n|\rho A|n\rangle \\
&= \mathrm{Trace}(\rho A)
\end{aligned}
$$

For such a normal quantum state f, the von Neumann entropy or *entanglement entropy* is defined as $S = -\sum_j p_j^2 \log p_j^2$.

I interpret the entanglement entropy as an (inverse) measure of the amount of information that the quantum system in a given state will yield through measurement. The larger the entropy of the quantum system, the less information can be extracted.

The measurement process cannot be performed by an external observer to elements within the interior, beyond the event horizon. We thus partition the event horizon of the black hole with elements each of area l_P^2, where l_P is the Planck length ($\cong 1.6 \times 10^{-35} m$) and assume the Planck area l_P^2 corresponds classically to the minimal projection $E_k H = [|y_k\rangle]$ of the pure vector state ω_{y_k}. Let N be the total finite number of partitions. By the 'no hair' hypothesis there is no preferred location on the event horizon, so that each partition element must have the same weighting. The entanglement entropy of this partition is thus given by;

$$S = -\sum_{1}^{N}\frac{1}{N}\log\left(\frac{1}{N}\right) = \log N = \log\frac{A}{l_p^2} = \log\frac{c^3A}{\hbar G}$$

with A the surface area of the black hole. This is similar to the Beckenstein–Hawking formula, given that log x is an increasing function of x. The logarithmic dependence guarantees that, for our formula, the entropy increases linearly with expansion of the phase space, as in the classical case.

This relates to general relativity via the macro-level Clausius definition of entropy S;

$$\oint\frac{\delta Q}{T} = \Delta S$$

where, integrated over a Carnot cycle, a change in energy $\delta Q = \delta T_{\mu\nu}$ results in a change in entropy ΔS corresponding to a change in space–time curvature. The Riemannian curvature tensor gives under contraction the Ricci curvature tensor $R_{\mu\nu}$, and its further contraction R, which between them capture the metrical structure of local space–time, This yields the Einstein field equations for gravity without the ad hoc cosmological constant $\Lambda g_{\mu\nu}$:

$$\frac{8\pi G}{c^4}T_{\mu\nu} = R_{\mu\nu} - \frac{1}{2}Rg_{\mu\nu}.$$

Thus, when a Carnot cycle is a good approximation to our quantum version of ergodic theory, our algebraic theory contains general relativity as an approximation.

4.6 Summary of chapter 4

I have, in chapter 4, building on the developments in chapters 1, 2 and 3, created a new approach to the understanding of Yang–Mills quantum fields. I have shown that such a quantum field can be identified with a local algebra by rolling up each such quantum field into a local quantum operator. The quantum field then becomes identified both as a section through the resultant fibre bundle and as an element of a local W*-algebra. We have also shown that, for such a construct, the relativistic quantum algebraic structure of the algebra can be identified, if G-type III, with the type III structure of the cross-product von Neumann algebra $(D) \times T$. This leads on to the development of quantum ergodic theory as a non-commutative extension of classical ergodic theory.

On this basis we are then, surprisingly, able to demonstrate that when this quantum ergodic theory is sufficiently close to the classical case to allow the application of a Carnot cycle to the entropy of the surface area of a black hole, then we have, embedded in our theory as an approximation, Einstein's field equations for gravity.

Further reading

Ashtekar A, Lewandowski J, Marolf D, Mourao J and Thiemann T 1995 Quantization of diffeomorphism invariant theories of connections with local degrees of freedom *J. Math. Phys.* **36** 6456–93

Wigner E, 1939 On unitary representations of the inhomogeneous Lorentz group *Ann. Math.* **40** 149–204

References

[1] Moffat J 1977 On groups of automorphisms of operator algebras *Proc. Camb. Philos. Soc.* **81** 237–43
[2] Moffat J 1973 Continuity of automorphic representations *Proc. Camb. Philos. Soc.* **74** 461–5
[3] Whitley R 1986 The Krein–Smulian theorem *Proc. Am. Math. Soc.* **97** 376–7
[4] Namioka I and Asplund E 1967 A geometric proof of Ryll–Nardzewski's fixed point theorem *Bull. Am. Math. Soc* **73** 443–5
[5] Moffat J 1974 Groups of automorphisms of operator algebras *PhD Thesis* (University of Newcastle-upon-Tyne)
[6] Salam A and Strathdee J 1974 Supergauge transformations *Nucl. Phys.* B **76** 477–82
[7] Kadison R and Ringrose J 1983 *Fundamentals of the Theory of Operator Algebras, Volume 1, Elementary Theory; Volume 2, Advanced Theory* (New York: Academic)
[8] Stormer E 1973 Automorphisms and equivalence in von Neumann algebras *Pac. J. Math.* **44** 371–83
[9] Hopf E 1932 Theory of measure and invariant integrals *Trans. Am. Math. Soc.* **34** 373–93
[10] Moffat J 1974 On groups of automorphisms of the tensor product of von Neumann algebras *Math. Scand.* **34** 226–30
[11] Moffat J, Oniga T and Wang C H-T 2017 Group Cohomology of the Translational Group and Diffeomorphism Invariant Quantum States(arXiv preprint: 1606.01578 v2 [math-ph]).
[12] Reed M and Simon B 1980 *Methods of Modern Mathematical Physics Volume 1. Functional Analysis* (New York: Academic)
[13] Pier J 1984 *Amenable Locally Compact Groups* (New York: Wiley)
[14] Segal I 1951 A class of operator algebras which are determined by groups *Duke Math. J.* **18** 221–65
[15] Hajian A and Kakutani S 1964 Weakly wandering sets and invariant measures *Trans. Am. Math. Soc.* **110** 136–51
[16] Dixmier J 1951 *Sur certains espaces considérés par MH Stone* (Brasília: Instituto Brasileiro de Educação)
[17] Takesaki M 1958 On the conjugate space of operator algebra *Tohoku Math. J.* **10** 194–203
[18] Takesaki M 1959 On the singularity of a positive linear functional on operator algebra *Proc. Jpn. Acad.* **35** 365–6

IOP Publishing

A Mathematically Coherent Quantum Gravity

James Moffat

Chapter 5

Implications

The Holy Quran enjoins us to reflect on the verities of Allah's created laws of nature[1].

5.1 Implications for mathematics

5.1.1 Clay Mathematics Institute Millennial question: Yang–Mills quantum theory and the mass gap

> *'The laws of quantum physics stand to the world of elementary particles in the way that Newton's laws of classical mechanics stand to the macroscopic world. Almost half a century ago, Yang and Mills introduced a remarkable new framework to describe elementary particles using structures that also occur in geometry. Quantum Yang–Mills theory is now the foundation of most of elementary particle theory, and its predictions have been tested at many experimental laboratories, but its mathematical foundation is still unclear. The successful use of Yang–Mills theory to describe the strong interactions of elementary particles depends on a subtle quantum mechanical property called the 'mass gap': the quantum particles have positive masses, even though the classical waves travel at the speed of light. This property has been discovered by physicists from experiment and confirmed by computer simulations, but it still has not been understood from a theoretical point of view. Progress in establishing the existence of the Yang–Mills theory and a mass gap will require the introduction of fundamental new ideas both in physics and in mathematics'.*

The description of Yang–Mills Theory with F-type symmetry breaking we have developed in chapter 4 is highly relevant to this question, as discussed in that

[1] From the Nobel Citation for Abdus Salam (winner of the Nobel Prize for Physics, 1979).

doi:10.1088/978-0-7503-2580-6ch5

chapter. At its core is the description of a quantum field as a section through a fibre bundle as used in proving a result I discussed in chapter 3 which I called at the time *A Remarkable Theorem*. The proof given there was guided by notions of physical intuition. Newton, for example, used his theory of fluxions to prove particular results, and then re-proved them using geometrical methods for public release. Knowing that the result was true made the result easier to prove formally. In this spirit, an alternative mathematical proof based on Mackey's theory of topological groups [1, 2] is now sketched. More detail is given in a paper which I published in 1977 [3]. I gladly acknowledge the input from Alain Connes through discussions.

A *character* acting on a locally compact abelian (l.c.a.) group G is a continuous group homomorphism from G to the unit circle. A result which will be useful to us is as follows.

If H is a closed subgroup of G and ρ is a character of H, then ρ extends to a character on G.

By exploiting Mackey theory, I gave an elegant proof of this result in [4]. In the remainder of this section I aim to prove the key result discussed earlier using only such mathematical methods.

Define α: $g \to \alpha_g$ to be a representation of the relativistic translational subgroup T of the Poincaré group as a gauge group of automorphisms of the local algebra $O(D)$. Provided this group representation α is weakly measurable, then it turns out that it is also norm, weakly and strongly continuous, and is implemented by a norm, weakly and strongly continuous unitary representation by unitaries in the algebra.

Proof. With these assumptions, we can conclude, as in the first part of the previous proof given in chapter 3 that the mapping α: $g \to \alpha_g$ is norm continuous and is implemented by a set $g \to W_g$ of commuting unitaries in the algebra.

Since $\|\alpha_g - i\| \geqslant \{\|\alpha_g(T)x - Tx\|; \quad \|T\| \leqslant 1\|x\| \leqslant 1\}$ it follows that the norm continuous mapping α: $g \to \alpha_g$ is strongly, and weakly, continuous. My foundational work at [3] then shows that we may choose the unitaries in a way that the mapping $g \to W_g$ is a Borel mapping from T to the unitary group of the local quantum field algebra $O(D)$ with the weak operator topology. Now, $O(D)$ is the fibre bundle consisting of all quantum fields defined over a compact subset of space–time. Let us define it to be the direct sum of the fibre algebras $\{\mathbf{A}(\boldsymbol{\xi}); \boldsymbol{\xi} \in D\}$ in the discrete case, during the Planck regime. In the continuous approximation beyond this initial regime as the Universe cools and expands, this yields an alternative representation of $O(D)$ as the *direct integral algebra* $O(D) = \int_{\mathrm{D}}^{\oplus} \mathbf{A}\,(\boldsymbol{\xi})\mathrm{d}\,\boldsymbol{\xi}$. The group representation α on $O(D)$ then defines a representation $\alpha(\xi)$ on each fibre algebra $\mathbf{A}(\boldsymbol{\xi}); \boldsymbol{\xi} \in D$, as I showed in [3], implemented by the set $g \to W_g(\xi)$ of commuting unitaries in the algebra $\mathbf{A}(\boldsymbol{\xi})$.

The unitary group of the centre of each algebra $\mathbf{A}(\boldsymbol{\xi})$ is isomorphic to the unit circle $\mathbf{O}$ in the complex plane since $\mathbf{A}(\boldsymbol{\xi})$ is a factor algebra. It is also easy to see that $\Gamma = \{\lambda W_g(\xi); \lambda \in \mathbf{O}, \text{and } g \in T\}$ is an abelian subgroup of the unitary group of $\mathbf{A}(\boldsymbol{\xi})$.

Define the mapping $\gamma(\lambda) = \lambda I$ which maps the unit circle $\mathbf{O}$ into the subgroup Γ, and the mapping $\eta(\lambda W_g) = g$ which maps the group Γ to the translation group T. The mapping η can be made well defined by factoring out its kernel. Thus the sequence;

$$0 \to \mathbf{O} \overset{\gamma}{\to} \Gamma \overset{\eta}{\to} T \to 0$$

is short exact, and Γ is an extension of $\mathbf{O}$ by T. We can identify the group Γ with the extension $\mathbf{O}\,\eta' T$ where; $\eta'(\lambda, g) = \eta(\lambda W_g(\xi)) = g$ via the mapping; $(\lambda, g) \to \lambda W_g(\xi)$.

Let the mapping J denote the identification $\mathbf{O} \times \mathrm{T} \leftrightarrow \Gamma \leftrightarrow \mathbf{O}\,\eta' T$ then J is a Borel mapping. Thus we see that the quantum field mapping $W(\xi)$: $g \to W_g(\xi)$ is a Borel measurable mapping from the translation subgroup T into the group Γ.

The group T is a separable l.c.a. group and is a standard Borel space. The unitary group of $\mathbf{A}(\boldsymbol{\xi})$ is also a standard Borel space, thus $W^{-1}(\xi)$: $W_g(\xi) \to g$ is a Borel mapping and η' is a Borel system of factors for $\mathbf{T}\,\eta' T$. There is thus, from Mackey theory, a locally compact topology on the group Γ relative to which it is a separable, l.c.a. group containing the unit circle $\mathbf{O}$ as a closed subgroup.

With this Mackey topology on Γ, we have that the identity map $\lambda \to \lambda$: $\mathbf{O} \to \mathbf{O}$ is a character on $\mathbf{O}$ and thus lifts to a character σ: $\Gamma \to \mathbf{O}$. This step in the proof is the equivalent to the devopment of the section mapping from a Stonean base space to a totally disconnected Stonean space in the previous proof.

For each unitary $W_g(\xi)$; $g \in T$ define $U_g(\xi) = \sigma(W_g(\xi)^* W_g(\xi)$. Then, since $\sigma|_O$ is the identity;

$$\sigma(U_g(\xi)) = \sigma(\sigma(W_g(\xi)))^* \sigma(W_g(\xi)) = \sigma(W_g(\xi))^* \sigma(W_g(\xi)) = 1 \;\forall\; g \in T.$$

Using the notation of my 1977 paper [3], we have that if $g, h \in T$ then $U_g(\xi) U_h(\xi) = \lambda(g, h) U_{gh}(\xi)$ for a 2- cocycle $\lambda(g, h) \in \mathbf{O}$. Since $U_g(\xi) \in \Gamma \;\forall\; g$, by definition of Γ, we can apply the group homomorphism σ to both sides of this relationship. Noting again that $\sigma|_O$ is the identity, this implies that

$$1 = \lambda(g, h) \text{ since } \sigma(U_g(\xi)) = 1 \text{ for all g.}$$

Thus all such 2-cocycles are trivial, and the mapping $g \to U_g$ is a unitary group representation.

We can identify the group Γ with the Borel system of factors $\mathbf{O}\,\eta' T$ and the mapping $W(\xi)$: $g \to W_g(\xi)$ is a Borel measurable mapping from the translation subgroup T into the group Γ. Hence the mapping $g \to \langle x, W_g(\xi)x\rangle$ is a measurable mapping for any x in the Hilbert space $H(\xi)$. In addition, the character σ: $\Gamma \to O$ is a continuous mapping, thus $\sigma \circ W(\xi)$: $g \to \sigma(W_g(\xi))$ is Borel. Combining these together it follows that the mapping $g \to \langle x, U_g(\xi)x\rangle = \sigma(W_g(\xi))^* \langle x, W_g(\xi)x\rangle$ is Borel measurable. Defining $U_g = \int_D^{\oplus} U_g(\xi) d\xi$, and similarly $W_g = \int_D^{\oplus} W_g(\xi) d\xi$, we have a Borel measurable representation $g \to U_g : T \to H$ implementing the automorphic representation $g \to \alpha_g$ on the Hilbert space H, where $H = \int_D^{\oplus} H(\xi)d\xi$. Thus, as in the previous proof, the mapping $g \to U_g$: is norm continuous. QED.

5.2 Further implications for Physics

5.2.1 Discrete closed strings in the early Universe

During the Planck regime of the very early Universe, we can model, as described in chapter 2, the evolution of space–time by the development of logical pathways through and between a set of evolving computational spin networks. From these beginning notions we are led to a discrete structure for space–time during the Planck regime corresponding to the first few atto- or possibly femto-seconds. As the Universe expands and cools this discrete structure develops into an emergent smooth four-dimensional manifold on which we can define particles (chapter 3) by exploiting the theory of heterotic closed strings. This leads to the following question:

Is it possible to describe an initial discrete form of a closed relativistic string during the Planck regime which then evolves into the form we know as a smooth approximation on a four-manifold?

To do this we consider a relativistic two-dimensional non-oriented chain of N 'atoms' arranged in a loop in the *light-cone reference frame* where the z-axis is boosted close to the speed of light. For such a reference frame the relativistic string coincides dynamically with its Newtonian approximation in the x–y plane. We assume that the atoms in this two-dimensional chain have a regular crystal spacing a between their equilibrium positions, and that these atoms are fairly tightly bound so that deviations from equilibrium due to external perturbations are small. The classical Hamiltonian for this system at time t is given by;

$$H(t) = \sum_{i=1}^{N} \frac{m}{2}\dot{x}_n(t)^2 + \frac{s}{2}(x_{n+1}(t) - x_n(t) - a)^2$$

where $x_n(t)$ is the location of the nth; 'atom' at time t and s is a 'spring constant' generated by the inter-atomic forces between the atoms of this regular, periodic structure. The spring constant does not change or decay over time. Note that these atoms represent clumps of mass, small enough such that the overall system behaviour is quantum in nature. We can consider such a structure as the discrete form of a closed string.

If we define $\bar{x}_n = na$ as the equilibrium position of the nth atom in the chain, then let $\varphi_n(t)$ be the deviation from this equilibrium position at time t. Thus;

$$\varphi_n(t) = x_n(t) - \bar{x}_n$$

$$\begin{aligned} \text{and } \varphi_{n+1}(t) - \varphi_n(t) &= x_{n+1}(t) - \bar{x}_{n+1} - (x_n(t) - \bar{x}_n) \\ &= x_{n+1}(t) - x_n(t) - (\bar{x}_{n+1} - \bar{x}_n) \\ &= x_{n+1}(t) - x_n(t) - a. \end{aligned}$$

Expressing the classical Hamiltonian as a function of the finite set $\{\varphi_n(t), n{=}1\ldots, N\}$ then yields;

$$H\{\varphi_n(t),\, n = 1\ldots,\, N\} = \sum_{i=1}^{N}\frac{m}{2}\dot{\varphi}_n(t)^2 + \frac{s}{2}(\varphi_{n+1}(t) - \varphi_n(t))^2.$$

We assume that the deviations $\varphi_n(t)$ from equilibrium are small. Mathematically we assume that;

$$|\varphi_n(t)| \ll a \text{ for all } n.$$

We also assume that in the Planck regime, quantum effects become important in understanding the behaviour of this loop. In the light cone reference frame, we can use the non-relativistic Schrödinger equation to describe the behaviour of the relativistic string in the x–y-plane orthogonal to the boost direction z. In order to make the transition from classical to quantum physics, we therefore replace the position and momentum values of the point atomic masses by $\hat{\varphi}_n$ and $\hat{p}_n$ for each of the atoms n in the loop, $1 \leqslant n \leqslant N$, transforming the classical functions into quantum operators. Each of the heterotic strings is closed, thus we identify the locations corresponding to the labels 1 and N. These operators are also functions of time, not shown explicitly to simplify notation. At a given point in time, they satisfy the following commutation relations;

$$[\hat{\varphi}_n, \hat{p}_m] = i\hbar\delta_{nm}$$
$$[\hat{\varphi}_n, \hat{\varphi}_m] = [\hat{p}_n, \hat{p}_m] = 0$$

where $\delta_{nm} = 1$ for $n = m$, otherwise $\delta_{nm} = 0$ and $\hbar = \frac{h}{2\pi}$ is Dirac's constant.

These commutation relations are the discrete equivalent of the commutation relations for a boson non-relativistic quantum field [5] and thus guarantee that the 'phonon' solutions are in fact bosonic quantum particles.

Making these operator substitutions in the classical Hamiltonian, it becomes the Hamiltonian quantum operator;

$$\hat{H}(\hat{p}, \hat{\varphi}) = \sum_{n=1}^{N}\frac{\hat{p}_n^2}{2m} + \frac{s}{2}(\hat{\varphi}_{n+1} - \hat{\varphi}_n)^2.$$

This turns out to be a good guess in terms of our black box model of a quantum system described in chapters 1 and 2. We now represent each of the location and momentum quantum operators $\hat{\varphi}_n$ and $\hat{p}_n$ as a discrete Fourier series relative to the harmonic basis vectors $u_n^k = \frac{1}{\sqrt{N}}\exp(ikan)$ where an is a measure of distance along the atomic ring, being equal to a count of the number of atoms times the inter-atomic distance. The coefficients of these basis functions are themselves quantum operators.

We have;

$$\hat{\varphi}_n(t) = \sum_k \hat{b}_k(t)u_n^k + \hat{b}_k^\dagger(t)u_n^{k*} \text{ where } u_n^{k*} = \frac{1}{\sqrt{N}}\exp(-ikan) = u_n^{-k}$$

$$\hat{p}_n(t) = \sum_k (-im\omega_k)\left(\hat{b}_k(t)u_n^k - \hat{b}_k^\dagger(t)u_n^{k*}\right).$$

These follow from quantum substitution into the equivalent expressions in the classical case, and correspond, as noted earlier, to a good guess.

The translational symmetry of our loop implies that $u_{N+n}^k = u_n^k$. This in turn implies that $\exp(ikaN) = 1$, thus $k = \frac{2\pi l}{aN}$ where l is an integer. Values of l in the range $-\frac{N}{2} \leqslant l \leqslant \frac{N}{2}$ are sufficient to resolve the discrete structure of the atomic lattice, thus k takes only the finite range of values $-\frac{\pi}{a} \leqslant k \leqslant \frac{\pi}{a}$.

Substituting these Fourier series into the quantum Hamiltonian $\hat{H}$ leads to the expression;

$$\hat{H} = \sum_k m\omega_k^2\left(\hat{b}_k\hat{b}_k^\dagger + \hat{b}_k^\dagger\hat{b}_k\right).$$

We can also derive from the commutation relations that $\hat{b}_k\hat{b}_k^\dagger = \hat{b}_k^\dagger\hat{b}_k + \frac{\hbar}{2m\omega_k}I$ with I the identity operator. Thus;

$$\hat{H} = \sum_k m\omega_k^2\left(\hat{b}_k\hat{b}_k^\dagger + \hat{b}_k^\dagger\hat{b}_k\right) = \sum_k m\omega_k^2\left(\hat{b}_k^\dagger\hat{b}_k + \hat{b}_k^\dagger\hat{b}_k + \frac{\hbar}{2m\omega_k}I\right)$$

$$= \sum_k 2m\omega_k^2\hat{b}_k^\dagger\hat{b}_k + \frac{\hbar\omega_k}{2}I.$$

This is the relativistic quantum Hamiltonian of a system of uncoupled quantum oscillators in the light cone reference frame, each of the form $\hat{H}_k = \hbar\omega_k(\hat{c}_k^\dagger\hat{c}_k + \frac{1}{2}I)$ with $\hat{c}_k = \sqrt{\frac{2m\omega_k}{\hbar}}\,\hat{b}_k$. Any energy level of the system thus corresponds to a discrete set of oscillator frequencies (equivalently oscillator energies) across various integer values of k. For a given value of k the energy level is determined by the expression; $|\, n_k\rangle = \frac{1}{\sqrt{n_k!}}(\hat{c}_k^\dagger)^{n_k}|0\rangle$ acting on the ground or vacuum state $|\, 0\rangle$ corresponding to n_k phonons of frequency ω_k. The operators $\hat{c}_k^\dagger$ are creation operators in the sense that the n_k-fold application of $\hat{c}_k^\dagger$ creates, from the vacuum state, these n_k phonons of frequency ω_k. The total energy of the system is thus defined in terms of a number n_k of such phonon particle states $\{n_k, \omega_k\}$ at a range of different quantised frequency levels ω_k. In this discrete lattice with translational symmetry, there is a natural upper bound or energy cut-off due to the finite range of values of k.

The discrete closed string thus acts like a stiff quasi-crystalline material and generates a finite set or 'stack' of bosonic particles at various quantised frequencies ω_k (equivalently energies $\hbar\omega_k$).

To symmetrise this as an approach equally applicable to fermions, we need to turn the commutators of the theory into anti-commutators.

5.2.2 The continuum limit beyond the Planck regime

When the total number N of atoms in our loop is very large, and the spacing between them becomes very small, a situation close to reality in most likely applications, it is possible to approximate the system by its continuum limit where $N \to \infty$ and $a \to 0$ while keeping the length of the loop (Na) the same. This limit is a non-relativistic quantum field with the phonon as the quantum of the field. As noted earlier, this also applies to the relativistic case where we view the system from the boosted *'light cone'* reference frame.

It turns out, as we will show, that the system energy in this limiting approximation, as the system cools beyond the Planck era, has the same characteristics as that of the discrete case, resulting in a phonon stack. This is exactly what would be expected from standard theory. The energy of the system is determined by a discrete set of 'bins' of phonon particle states $\{n_k, \omega_k\}$ each consisting of a number n_k of phonons at the quantised frequency level ω_k. Thus the same system concept still applies even in the continuum approximation of very large numbers of 'atoms' in a dense ring. However, some form of imposed energy cut-off is required in this case to keep the number of 'bins' finite and thus avoid the assumption of infinitely large energies in the system.

5.2.3 The continuum limit as a quantum field

Let us assume that N, the number of atoms in the linear chain, is in fact very large, and that the deviation of the nth atom from its equilibrium state, $\varphi_n(t)$, is well approximated by a continuous function $\varphi(x, t)$ of length x along the chain.

In this continuous approximation, which holds beyond the Planck regime, we have,

$$\varphi_n(t) \to \varphi(na, t) = \varphi(x, t)|_{x=na}$$
$$\frac{\varphi_{n+1}(t) - \varphi_n(t)}{a} \to \frac{\partial\varphi(x, t)}{\partial x}.$$

We turn this into a non-relativistic quantum field through a standard process known as canonical field quantisation, in which the dynamics of the field become relationships between quantum operators. In our case, since we are dealing with a relativistic regime in the light cone reference frame, boosted along the z-axis, the dynamics of the two-dimensional string in the x–y plane are represented by the Schrödinger equation;

$$i\hbar\frac{\partial\psi}{\partial t} = \hat{H}\psi$$

where for our application, the discrete Hamiltonian operator is given by;

$$\hat{H}(\hat{p}, \hat{\varphi}) = \sum_{n=1}^{N} \frac{\hat{p}_n^2}{2m} + \frac{s}{2}(\hat{\varphi}_{n+1} - \hat{\varphi}_n)^2.$$

In terms of our black box model of quantum theory, this also turns out to be a good guess. We first consider the continuum limit of this Hamiltonian operator. The total area enclosed by integrating the continuous function $\varphi(x, t)$ over the length L of the chain, with $L = Na$, is given by $\int_0^L \varphi(x, t)dx$. This is equivalent in the discrete case to $a\sum_{i=1}^{N}\varphi_n(t)$. Making these limiting approximations we have, for the classical Hamiltonian;

$$\begin{aligned} H(\varphi_n(t)) &= \sum_{i=1}^{N} \frac{m}{2}\dot{\varphi}_n(t)^2 + \frac{s}{2}(\varphi_{n+1}(t) - \varphi_n(t))^2 \\ \rightarrow H(\varphi(x, t)) &= \frac{1}{a}\int_0^L \frac{m}{2}\dot{\varphi}(x, t)^2 + \frac{s}{2}\left(a\frac{\partial\varphi(x, t)}{\partial x}\right)^2 dx. \end{aligned} \tag{5.1}$$

In the continuum limit we can also approximate the chain of discrete masses of value m separated by regular spacings of length a, by a local mass density ρ at each point x, with $\rho = \frac{m}{a}$ and a spring constant per unit length $s \rightarrow \frac{s}{a}$. Inserting these into equation (5.1) above we have;

$$\begin{aligned} H(\varphi(x, t)) &= \int_0^L \frac{\rho}{2}\dot{\varphi}(x, t)^2 + \frac{s}{2}\left(\frac{\partial\varphi(x, t)}{\partial x}\right)^2 dx \\ &= \int_0^L \frac{(\rho\dot{\varphi}(x, t))^2}{2\rho} + \frac{s}{2}\left(\frac{\partial\varphi(x, t)}{\partial x}\right)^2 dx \\ &= \int_0^L \frac{(p(x, t))^2}{2\rho} + \frac{s}{2}\left(\frac{\partial\varphi(x, t)}{\partial x}\right)^2 dx \end{aligned}$$

where $p(x,t)$ is the local momentum at the location x at time t. Through the process of canonical field quantisation, the Hamiltonian itself now becomes a quantum field operator. First we define the *Hamiltonian density operator HD*, which depends on $\hat{p}(x, t)$ and $\hat{\varphi}(x, t)$ as follows;

$$HD(\hat{p}, \hat{\varphi}) = \frac{(\hat{p}(x, t))^2}{2\rho} + \frac{s}{2}\left(\frac{\partial\hat{\varphi}(x, t)}{\partial x}\right)^2.$$

This process is sometimes called 'second quantisation' because we go firstly from the classical field to the Schrödinger equation and then secondly to the Hamiltonian density operator *HD*. For such a quantum field theory in which the dynamics are derived from the Schrödinger equation, the Hamiltonian of the field, Ξ is given by;

$$\Xi(\hat{p}, \hat{\varphi}) = \int dx HD(\hat{p}(x, t), \hat{\varphi}(x, t)) = \int dx \ \hat{\varphi}^{\dagger}(x, t)\hat{H}(\hat{\varphi}(x, t))\hat{\varphi}(x, t)$$

where $\hat{\varphi}^{\dagger}$ is the operator adjoint of $\hat{\varphi}$. This is derived by exploiting the assumption that the corresponding Lagrangian density satisfies the Euler–Lagrange equation (an 'on-shell' solution). In some texts it is taken simply as a definition of the field Hamiltonian for a non-relativistic field.

5.2.3.1 The commutation relations of the field

The canonical field quantisation process leads to the operator fields $\varphi(x, t) \to \hat{\varphi}(x, t)$ and $p(x, t) \to \hat{p}(x, t) = i\hbar\hat{\varphi}^{\dagger}(x, t)$. We can derive commutation relations for these on the basis of the commutation relations for the discrete case, which are;

$$[\hat{\varphi}_n, \hat{p}_m] = i\hbar\delta_{nm}$$
$$[\hat{\varphi}_n, \hat{\varphi}_m] = [\hat{p}_n, \hat{p}_m] = 0.$$

Taking the continuum limit, denoted →, and then applying the canonical field quantisation approach these become;

$$[\hat{\varphi}_n, \hat{p}_m] = i\hbar\delta_{nm} \to [\hat{\varphi}(x, t), \hat{\varphi}^{\dagger}(x', t)] = \delta(x - x')$$
$$[\hat{\varphi}_n, \hat{\varphi}_m] = 0 \to [\hat{\varphi}(x, t), \hat{\varphi}(x', t)] = 0$$
$$[\hat{p}_n, \hat{p}_m] = 0 \to [\hat{\varphi}^{\dagger}(x, t), \hat{\varphi}^{\dagger}(x', t)] = 0$$

where $\delta(x - x')$ is the Dirac delta function.

These commutation relations characterise a *boson* quantum field.

5.2.3.2 Energy eigenstates

Since the Schrödinger Hamiltonian $\hat{H}$ is (by definition) itself a Hermitian operator with real energy eigenvalues, its eigenstates $u_k(x, t)$ form a basis of the Hilbert space $L^2(X, \mu)$ of square-integrable wave functions on which $\hat{H}$ acts, for some Lebesgue measure space X which is a subset of space–time. In our case the potential function does not depend on time, but only on location since the spring constant s does not change over time. For this case we can define the energy eigenstates as $u_k(x)$; these are functions only of location, x.

5.2.3.3 Representing the abstract field operators

We now define the following representation of the abstract field operators;

$\hat{\varphi}(x, t) = \sum_k \hat{a}_k(t)u_k(x)$ with all of the time dependence taken by the operators $\hat{a}_k(t)$

From this we also have $\hat{p}(x, t) = i\hbar\hat{\varphi}^{\dagger}(x, t) = i\hbar\sum_k \hat{a}^{\dagger}{}_k(t)u^*{}_k(x)$

where * denotes complex conjugate.

5.2.3.4 The phonon stack in the continuum limit

It follows that the quantum field Hamiltonian Ξ is given by;

$$\begin{aligned}
\Xi(\hat{p}, \hat{\varphi}) &= \int \hat{\varphi}^{\dagger}(x, t)\hat{H}(\varphi(x, t))\hat{\varphi}(x, t)dx \\
&= \int \sum_{kk'} \hat{a}_k^{\dagger}(t)u_k^*(x)\hat{H}(\varphi(x, t))\hat{a}k'(t)u_{k'}(x)dx \\
&= \int \sum_{kk'} \hat{a}_k^{\dagger}(t)u_k^*(x)\hat{H}(\varphi(x, t))u_{k'}(x)\hat{a}_{k'}(t)dx \\
&\quad \text{(since } u_{k'}(x) \text{ is just a complex valued function)} \\
&= \sum_{kk'} \hat{a}_k^{\dagger}(t)\hat{a}_{k'}(t) \int u_k^*(x)\hat{H}(\varphi(x, t))u_{k'}(x)dx \\
&\quad \text{(since } \hat{a}_k^{\dagger} \text{ and } \hat{a}_{k'} \text{ are not functions of } x) \\
&= \sum_{kk'} \hat{a}_k^{\dagger}(t)\hat{a}_{k'}(t) \int u_k^*(x)\varepsilon_{k'}u_{k'}(x)dx \\
&\quad \text{(where } \varepsilon_{k'} \text{ is the energy eigenvalue for the eigenstate } u_{k'}(x)) \\
&= \sum_{kk'} \hat{a}_k^{\dagger}(t)\hat{a}_{k'}(t)\varepsilon_{k'}\delta_{kk'} \\
&= \sum_{k} \hat{a}_k^{\dagger}(t)\hat{a}_k(t)\varepsilon_k.
\end{aligned}$$

It is thus composed of a stack of phonon particle states $|n_k\rangle$ at various discrete frequencies ω_k corresponding to the quantum energy levels ε_k as in the discrete case.

However, there is one key difference. In the continuum limiting case, there is no obvious energy cut-off as in the discrete lattice case with translational symmetry. In fact we need to either impose such a cut-off or else renormalise the energy relative to some starting value to avoid this infinity.

In summary, from finite beginnings a closed string can be interpreted as converging, in the continuum limit, to the version we are more familiar with as we deal with a universe which is cooler larger and well beyond the Planck regime.

We complete this book with a brief non-technical overview of our algebraic form of reality; a form which Einstein could not reach.

5.2.4 A brief non-technical overview

At the beginning of the Big Bang Universe, Einstein's theory of the gravitational field (general relativity) breaks down since it implies that the curvature is infinite, which does not really make sense. So other things must be having an effect—things which mattered when the Universe was very small, very hot, very dense, and very curved. What we need is the theory of quantum mechanics, which tells us how Nature works when distances and objects are very small. Bringing quantum mechanics to bear on the origin of the Universe, and unifying it with very curved space–time, is a problem which keeps a number of us awake at night. In this book I identify the key issue as being that, like a less favoured aunt, general relativity brings with it far too much baggage. We redefine the underlying notions of space and time as sparse properties of a discrete quantum theory. This we do using ideas drawn from the logic of quantum computation. A trajectory in space–time becomes a quantum algorithm corresponding to a set of logical steps across a network of entangled spin states—a process known in quantum computing as quantum teleportation. In this way I produce a simpler, discrete set of notions about gravity. Gravity is a set of waymarks which tell you how to navigate the network. These ideas are then compatible with our most modern definition of a quantum state as a normalised, positive element of the dual space of an algebra of quantum observables. This is the essence of chapters 1 and 2. In chapter 3, I show how particles emerge in such a theory by exploiting the mathematics of heterotic string theory projected from 10 dimensions down to our four-dimensional space–time. Chapter 4 is the final development of our theory. Here I create a new physics and mathematics corresponding to a new way of thinking about what I call local algebras. These are algebras of quantum fields defined over a compact subset of space–time and subject to the constraints of symmetry and symmetry-breaking processes. I create a quantum version of ergodic theory and combine it with a quantum version of entropy called 'entanglement entropy' in quantum computation.

In the final part of chapter 4, I show that when the quantum entropy of a system is close to its classical approximation, then Einstein's field equations for gravity (i.e. general relativity) emerge as an approximation based on the application of the Carnot cycle to the surface area of a black hole. This was a surprise to me, and a most satisfying conclusion.

References

[1] Mackey G 1957 Borel structure in groups and their duals *Trans. Am. Math. Soc.* **85** 134–65

[2] Mackey G 1957 Ensembles Boreliens et Extensions des Groupes *J. Math. Pure Appl.* **36** 171–8

[3] Moffat J 1977 On groups of automorphisms of operator algebras *Proc. Camb. Philos. Soc.* **81** 237–43

[4] Moffat J 1974 Groups of automorphisms of operator algebras *PhD Thesis* Newcastle University, UK

[5] Greiner W and Reinhardt J 1996 *Field Quantisation* (Berlin: Springer)

Lightning Source UK Ltd.
Milton Keynes UK
UKHW051926131120
373332UK00003B/62

9 780750 325783